URBAN CLIMATE CHANGE CROSSROADS

Urban Climate Change Crossroads

Edited by

RICHARD PLUNZ and MARIA PAOLA SUTTO
Columbia University, USA

ASHGATE

QC
981.7
.U7
U73
2010

Original edition printed 2008

Richard Plunz and Maria Paola Sutto have asserted their right under the Copyright, Designs and Patents Act, 1988, to be identified as the editors of this work.

Published by
Ashgate Publishing Limited
Wey Court East
Union Road
Farnham
Surrey, GU9 7PT
England

Ashgate Publishing Company
Suite 420
101 Cherry Street
Burlington
VT 05401-4405
USA

www.ashgate.com

This publication is issued in conjunction with the forum *Urban Climate Change at the Crossroads* held at the Camera di Commercio di Roma, Sala delle Conferenze, Via de' Burro' 147, Rome, Italy, February 4-5, 2008. Forum Working Committee: Cinzia Abbate, Francesca Billi, Flaminia Gennari, Bartolomeo Pietromarchi, Richard Plunz, Maria Paola Sutto, Giuseppe Tripaldi. The Forum and Publication have been made possible with the support of the Camera di Commercio Industria Artigianato e Agricoltura di Roma - Ambiente e Territorio (AeT); the Fondazione Adriano Olivetti in Rome; the Urban Design Lab of the Earth Institute at Columbia University in New York; the Italian Ministry of the Environment in Rome, and the Italian Trade Commission in New York.

Designed by Dimitrios Vlachopoulos.

British Library Cataloguing in Publication Data
Urban climate change crossroads.
   1. Urban climatology--Congresses. 2. Climatic changes--
   Environmental aspects--Congresses. 3. Climatic changes--
   Risk management--Congresses.
   I. Plunz, Richard. II. Sutto, Maria Paola.
   363.7'3874'091732-dc22

Library of Congress Cataloging-in-Publication Data
Plunz, Richard.
   Urban climate change crossroads / by Richard Plunz and Maria Paola Sutto.
      p. cm.
   Includes bibliographical references and index.
   ISBN 978-0-7546-7999-8 (hardback) -- ISBN 978-1-4094-0078-3 (pbk.) -- ISBN 978-0-7546-9980-4 (ebook)
1. Urban climatology. 2. Climatic changes. 3. Urban heat island. 4. Urban ecology (Sociology) I. Sutto, Maria Paola. II. Title.
   QC981.7.U7P57 2009
   551.65173'2--dc22

2009035480

ISBN 978 0 7546 7999 8 (hbk)
ISBN 978 1 4094 0078 3 (pbk)
ISBN 978 0 7546 9980 4 (ebk)

Mixed Sources
Product group from well-managed
forests and other controlled sources
www.fsc.org Cert no. SA-COC-1565
© 1996 Forest Stewardship Council
FSC

Printed and bound in Great Britain by
MPG Books Group, UK

Contents

"To build is to collaborate with earth, to put a human mark upon a landscape, modifying it forever thereby; the process also contributes to that slow change which makes up the history of cities."

Marguerite Yourcenar, *Memoirs of Hadrian,* 1951

In Rome, on February 4 and 5, 2008, within the formidable walls of the ancient Temple of Hadrian (145ad), a unique group of individuals assembled for a forum entitled "Urban Climate Change at the Crossroads." Indeed, in that historical setting, it was inevitable to reflect on the "end of history," particularly poignant relative to the subject at hand. This extraordinary volume is an outcome of the event, organized by the Camera di Commercio Industria Artigianato e Agricoltura di Roma - Ambiente e Territorio (AeT); the Fondazione Adriano Olivetti in Rome; and the Urban Design Lab of the Earth Institute at Columbia University in New York. We are very grateful for the support of the above collaborators; as well as the Italian Ministry of the Environment, and the Italian Trade Commission in New York, which provided additional support specifically for the preparation of this publication. At the AeT, we are especially grateful to its Director, Giuseppe Tripaldi, who has enthusiastically embraced the project since its inception and has provided continuous support throughout; and to Alessandra Nutta and Stefano Rossi Crespi for the operational and careful support work for the forum; we are also grateful to Cinzia Abbate who served as the Italian coordinator for the project. At the Fondazione Adriano Olivetti, we are indebted to Laura Olivetti, President, for her continuing interest; to Flaminia Gennari and Bartolomeo Pietromarchi for their initial contributions to the conference content; and to Francesca Billi, who served as coordinator for their sponsorship. At the Earth Institute, we are indebted to Steven A. Cohen, Executive Director, who early on provided encouragement and support. We also acknowledge the support of Corrado Clini, General Director of the Italian Ministry for the Environment in Rome, and of Claire Servini of the Italian Trade Commission in New York. As well, Cynthia Rosenzweig of the NASA Goddard Institute for Space Studies, Steven Hammer of the Urban Climate Change Research Network, and Jeanette Limondjian, of Barnes & Noble Booksellers were all most generous with their time and suggestions about the conference participants and format. For their assistance in the complicated task of assembling the material in this volume, we wish to thank Gabriella Folino, Merrell Hambleton, Monika Kowalczykowski, and Emily Weidenhof for their general editorial work; and Suzanne Fass for copyediting. Most important of all, we are deeply indebted to Dimitris Vlachopoulos of the Urban Design Lab, who made the graphic design and production. Without his extraordinary skills, this record could not exist.

Richard Plunz, Maria Paola Sutto

This volume is a remarkable collection of ideas, visions, and tools for rescuing a global civilization on the path to calamity. Urban climate change is a crossroads in two very different senses. One is historical. With the world now more than half urban, and given the ecological consequences of the world's high-consumption urban centers, we are at an ecological crossroad. We either head off the worst of ecological collapse through concerted and forward-looking action, or we face a "Mad Max future" of dystopia, violence, and upheaval. The second crossroad is intellectual. Our individual disciplines are unable to grasp the magnitude of the economic-ecological challenges ahead. For that we need to work holistically, calling on the knowledge of climatologists, engineers, sociologists, economists, public health specialist, designers, architects, community organizers, and more. The intellectual crossroad is nothing less than a new intellectual field of Sustainable Development. This book makes vivid the first crossroad by achieving the second.

The range of insights is enormous. Richard Plunz sets the stage by reminding us of the Club of Rome's dramatic entry onto the world's consciousness in 1972. *Limits to Growth* was dismissed by the mainstream then, but its warnings have proved to be inescapable and sometimes eerily on track, such as the estimate of atmospheric carbon concentration for the year 2000. We are not facing an inevitable collapse. There is still time to head off the worst, and to adapt to the changes and dislocations that are already in train, but only if we mobilize our knowledge and our politics to do so. Perhaps the current financial maelstrom will have a silver lining, reminding us of our civilization's fragility and our vulnerability to the risks of abrupt change that Plunz emphasizes. The volume's remaining essays give us insights as to how that can be accomplished.

There are no finished proofs or theorems in this volume, only the outlines of a new edifice of sustainable urbanization. Sustainable cities will be sustainably integrated into local and global ecological processes, using energy-efficient and low-carbon technologies, emphasizing the quality of life in local communities, and building healthy systems for transport, recreation, and employment. Ecology and economics will merge, rather than being distant and often conflictive modes of thinking and action. Design in its broadest sense – integrating technology, spatial analysis, community participation, and a deep appreciation of ecology – will provide an integrative framework for action.

The essays address these challenges in such diverse and interesting ways that they hint at the emerging field of Urban Sustainable Development, but they do not yet define the field. It's at this point simply too vast, too interconnected, and too incipient in problem solving to have a clear definition and set of tools. The essays take the same challenge – climate change mitigation and adaptation – and address it through a remarkable multiplicity of perspectives. How should communities govern themselves to achieve true sustainability? What do we mean by environmental justice and how can we achieve it? What forms of communication can help the public to understand the deep, complex, and highly uncertain sciences and engineering related to climate change? How can design mobilize diverse communities of expertise to protect cities? Some of the essays are deeply practical and in the present, such as Cynthia Rosenzweig's description of the sustainability initiatives of New York City, in which she plays an important leadership role. Other essays are highly speculative, looking forward to the disastrous or successful outcomes that await us depending on how we traverse these historical crossroads.

Readers will be stimulated, perplexed, and challenged throughout. They will take a new energy and commitment from the volume. And most important, they will understand that the task of sustainable urban development is one of the world's greatest challenges, with our generation at a world-sharing crossroad.

Jeffrey D. Sachs
Director, The Earth Institute

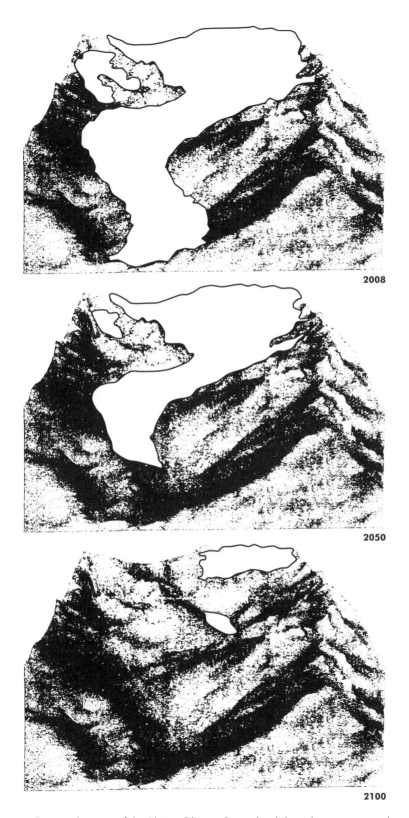

2008

2050

2100

Projected retreat of the Rhône Glacier, Switzerland, based on a numerical model developed by the Laboratory of Hydraulics, Hydrology, and Glaciology (EPFL) at the ETH, Zurich. http:iacs.epfl.ch/. The Rhône Glacier is the source of the Rhône River and a primary source of Lake Geneva.

CHAPTER 1:   **The Design Equation**
Richard Plunz

In reflecting on the Club of Rome and their initial 1972 publication, *Limits to Growth: A Report to the Club of Rome Project on the Predicament of Mankind*, one is still confronted by a very relevant document in terms of understanding the complexity of our present condition.[1] *Limits to Growth,* and later related studies, summarized for a whole generation a deep concern about the implications inherent to the society of consumption that was in evolution by the '60s. Indeed, especially in the United States, consumption was promoted as a new economic order in the post-war global reconfiguration. We tend to forget that the same debates today are far from new - only the immediacy is growing. For example, the Club of Rome predictions about CO_2 concentration were on the mark. (Fig. 1.1)

Or consider the work of the British economist, Ezra Mishan, who was also associated with the Club of Rome - and who still provides a relevant critical perspective on the culture of consumption. Especially useful is his "Mishan Model," published in *The Costs of Economic Growth* in 1967, that effectively demonstrates the limits of automobile culture.[2] Mishan's crucial judgment was that even after public consensus would one day conclude that fewer or no automobiles is socially preferable to transport dysfunction, the system will not have built-in controls such that fewer automobiles can ever be obtained without extraordinary intervention external to the transport sector. And using the automobile as a metaphor for our global growth in general, we can begin to understand the enormity of the questions surrounding global environmental change today - and that the growth of global consumption and environmental degradation

---

Marianella Sclavi

I believe that one of this meeting's aims is consideration of how to take a step forward toward a positive scenario, and it must be a scenario with a multi-track approach. Our goal must be a scenario that goes beyond linear priorities, and follows a win-win, or better, a mutual gains kind of approach: all the problems to be tackled, all **Our goal must** considered at the same time. It may be easier to tackle **be a scenario** low-cost housing together with more jobs and with sta- **that goes** bilization of the climate than confronting each of these **beyond linear** problems one at the time. In the '60s it was called a "vir- **priorities, and** tuous circle" approach as opposed to a "vicious circle." **follows a win-** We need many examples of best practices - showing **win, or better,** what has been feasible in other parts of the world, and **a mutual** **gains kind of** what Albert Hirschman called a "bias toward hope." * **approach**

is not easily reversed without unprecedented global intervention.

Mishan provided a paradigm for the "manifest destiny" of the so-called American Century, as global power has played itself out. Until very recently, in so many quarters, it was the American hegemony that was expected to govern the planetary trajectory. In the run-up to the U.S. invasion of Iraq in 2004, President George W. Bush used the argument that there must be ". . . no holding back, no compromise, no hesitation" in protecting the American "way-of-life." [3] But the evidence is well in place that the American way of life is itself not sustainable, put in evidence by the futility of the war itself and its connection to the growing depletion of the world's oil reserves that remain the crucial life-support for the culture of consumption in the United States. Now the "American way-of-life" enters an excruciatingly difficult period, but ultimately for the better. We know that in absolute terms, within recorded history, there is no precedent on a global scale for the environmental interventions that now unfold. But in terms of relative scale - of human response - that's another very interesting question altogether. We are not alone in history. And in history there have been powerful precedents.

Among my favorite places in the world are the sites of the classical Greek cities of the southwest Aegean shore. Priene, in particular, is a place where one is invited to contemplate so many things about our present environmental dilemmas - and our human frailties, upheavals, struggles. When one looks out from its promontory across so many kilometers of green agricultural fields, where there was once as many kilometers of sea, one can't help but reflect on those souls who lived there millennia ago - and in the nearby cities of Miletus, Ephesus, and Heraclea - and at what point those city-states must have felt the enormity of their ecological miscalculations, and the destruction of their worlds.[4] They were too late in realizing that

---

* New Haven: Yale University Press, 1971.

Domenico Cecchini
New types of emissions maps would entail a complex, variegated result, consisting of different parts which gradually we would try to connect and link up, but still along the same lines of this necessity - an ethical necessity let's say - in the end, an urgent need to act even if there is so much theoretical pessimism.

Luigi Massa
It's very complicated to believe that adult public consensus in our country is really able to change our social model. Our model has been strongly targeted towards consumerism for forty years: a reversal of this trend

Our model has been strongly targeted towards consumerism for forty years

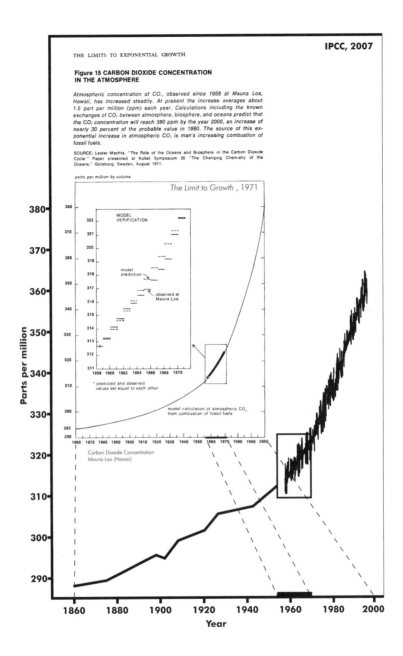

Fig1.1. Superimposition of CO_2 atmospheric concentrations:
The Limit to Growth, 1971: 1860-1960 Data, 1960-2000 Projections
IPCC Fourth Assessment Report, 2007: 1860-2000 Actual Data

the deforestation that provided the material for their growth would also be the source of their destruction. One supposes that those citizens were also obliged, like ourselves, to initiate forums like this one.

The Meander River silted over the harbors and the livelihoods of the cities that it had formerly nourished - in a painfully slow and unremitting process. Strabo documents how the citizens of Priene had already lost their port by his time (63bc–24ad) - it was 40 stadia distant. In fact we know that Priene was moved to its present location in the 4th century bc, with the ruins of the earlier city still remaining today at some unknown location under the silt. I mention this because I think that understanding the adaptation question engages a huge historical dimension - one can even say that environmental adaptation has been a "normal" condition of human history - but one that is anathema to modern design culture.[5] We simply do not look back, and relative to this history, and especially to ecology, it is possible that we are now in a "post-historical period" when the easiness of the past is no longer so resonant? The brilliance of Bill McKibben's book *The End of Nature* was in articulating precisely this point, more than a decade ago.[6]

The gamble for ecological survival has always been reliant on technology and design - and when the technological limits are obvious, the design adaptation has to be made. As with the cities of the Meander, when design adaptation is inadequate, humankind has moved on. The design imperative appeared with Katrina in New Orleans. Now New York faces a moment of truth with hurricanes and sea level rise. So does Bangkok, which is sinking as the sea rises. So does Quito, which is losing its water supply as the glaciers melt (fifteen years out). And so on. We have definitively arrived at the issue of "limits" - questions anticipated by the Club of Rome and

---

would create mistrust. We should really jump a generation ahead, because we have to work on the next decision makers; we have to act on educators; and we have to work on children. Today, as far as Italy is concerned, only the European Commission will have the capacity to force the decision makers, by linking structural funds to good environmental behaviors. Change will come when we will be no longer using gross domestic product (GDP) as a measure of economic success; but also measuring, for example, the reduction of carbon dioxide emissions. Reversal of such practice has to lead to better decisions by policy-makers.

Domenico Cecchini
I got inspired by Richard's presentation: let us imagine

We have to work on the next decision makers; we have to act on educators; and we have to work on children

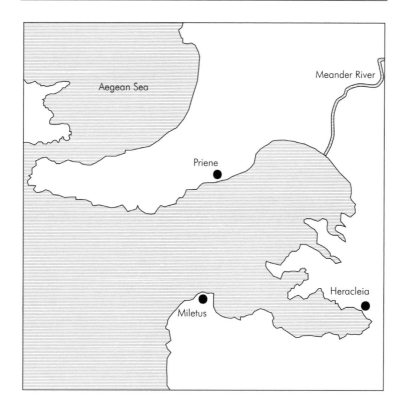

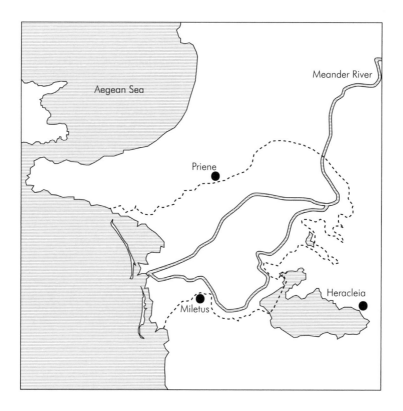

Fig 1.2. Silting of the Meander River

now far more immediate.

How urban adaptation occurs goes beyond building seawalls; moving from flood plains; inventing more robust infrastructures; and the like. A part of our consideration on the urbanism side is the changing nature of urban enterprise itself - of industry. Urban climate change adaptation involves urban economies - and not just global, but local economies - new forms of production, and specifically new urban industry that can integrate with the adaptation process itself. This trajectory engages what has been popularly described as a new field of "restoration ecology" and is corollary with degraded urban contexts - and therefore has the potential to generate a new urban production.

In cities, the restoration economy is increasingly critical, as restoration science moves from built fabric to the urban "natural environment" following the logic that degraded urban ecosystems may actually provide more economic opportunities than detriments.[7] Involved are emerging new forms of "natural" resources, including the "spent nature" of the brown fields. The concept of nature as "accumulation strategy" is expanding to include the redefinition of nature itself. Degraded and consumed environments are no longer seen as a liability but rather as a powerful new frontier of capital accumulation - such that "nature is increasingly if selectively replicated as its own marketplace."[8] The social and monetary costs and benefits must factor this expanded arena of considerations, and this economic adaptation must go hand in hand with physical adaptation within climate strategies.

Embedded in the concept of the new urban production is a critique of our inherited premises of "economic growth." The old models rely on the assumptions of always-expanding material consump-

---

that we in this room are in the city of Troy 2,000 years ago, discussing things that are about to happen. Cassandra emerges from the walls of Troy shouting, "The city is burning, wake up!" Let us imagine we are Trojans trying to avoid disaster - so optimism in practice and pessimism in theory.

Harriett Bulkeley
How do we draw on the experience of what has already happened? How do we know when a development has been successful, what have been its impacts? What I think is missing is the question: how do we learn from mistakes? How do we learn from bad practice without stigmatizing people and making it seem there has been failure, when actually they have just been learning? No-

How do we learn from bad practice without stigmatizing people?

tion fueled by always-growing economies. Also characteristic is a separation of economics from ecology, with economists ignoring natural contexts, and ecologists ignoring human contexts – a formulation that no longer guarantees human progress. Urban economies that thrive in the new era will have to function within a second modernity, the antithesis of the first that rejects the "canonical assumptions" of insatiable wants and infinite resources leading to supposed "growth forever."[9] Here we engage the question of a new beginning, and of climate change as a positive economic force. A positive prognosis is possible, especially if, as the German sociologist Ulrich Beck argues, now is a moment of a "second modernity" which is "being born within the interstices of the first modernity, most of all within its cities. . . ."[10]

When science interacts with the city, design is an essential catalyst, and geo-environmental analysis aside, the primary language for urban intervention has entailed "design," deployed as a catalytic urban tool and an effective means of visualizing for public discourse the complexities of the intervention strategies that urban design research must evoke.[11] The huge question is whether or not the design fields are at all adequately equipped to manage the adaptation problems: Beck's "glocalization" challenge; the new technological realities; the post-historical context in general. Certainly the context for design activity is changing rapidly, but Western design culture has evolved over the past four or five centuries such that appearances aside, a change in trajectory on the scale of the ecological change that we now face will be challenging, to say the least.

Design is intimately connected to both the ecological and political science realms - and ultimately to the question of power - and abuse of power. Yet the world of design culture has always resisted forthright admission of this reality - and for the "design equation"

Fig 1.3. Homepage for Design Competition for Emergency Housing

within the climate change challenge, I see this denial as complicit to the incapacity of the design world to deal with the environmental challenges which climate change is intensifying.

It is especially during the past 150 years or so that our present professional limitations have crystallized: with the separation of architecture from engineering; the evolution of isolated design discourse and pedagogy, with artificial distinctions between design of buildings, urbanism, landscape, interiors, products, and the like. These old problem sets are now obsolescent, with redundancies in terms of intellectual and operational outlook. Today, especially for environmental challenges related to global development and sustainability, and for questions of adaptation by design, there is a growing challenge to these institutionalized categories, not only from the changing nature of the problems, but also from new shared technologies of representation and production.

For decades architecture had considered itself the "mother" activity for design of the built environment - even as it became more and more marginalized. During the present wave of global urbanization this trend has intensified. The field has always accounted for a small percentage of the built environment, but in the new scale of global urbanization, it is minuscule. High architectural discourse has become more and more restricted to the world of high fashion, to the exclusion of the other some 99 percent of building. The problem is that high fashion is emblematic of the "mainstream" values of design culture - and a new legitimacy for design can not be achieved simply through manifesto. There is a need for new fundamental design knowledge.

While the demand for built environment expertise is growing, the building industry world-wide is said to be the most extraordinarily

---

body really wants to be named as "bad practice."

Luigi Massa
Human beings have a hard time metabolizing negative experiences and disasters. As an observer who wrote about a very dramatic event in human history, Pliny the Elder described the disaster of Pompei and Herculaneum in a very immediate and remarkable way. Yet we have rebuilt, and we have rebuilt much closer to the volcano than our forebears ever did, despite the fact that the story of the eruption survives through the writings of Pliny the Elder.

Harriett Bulkeley
It is important to mention that when we are dissecting

Human beings have a hard time metabolizing negative experiences and disasters

wasteful and destructive of all human enterprise, placing it at the epicenter of any adaptation strategy. Of course the design market-place itself is adapting to the new global context - including academia. But it is quite possible that there is far more interest in design culture from the outside looking in, than vice versa. As the natural sciences "urbanize" -and they certainly are in our present wave of "glocalization" - they are finding that urban science gets complicated, principally through the necessity for political engagement - simply stated, the necessity for "design" visualization. I might add that climate science is no exception - it, too, is urbanizing.

It is fortunate that science and design are being coerced into new relationships by new realities. The kind of creative process inherent to design methodologies, for so long anathema to the scientific "method," is suddenly gaining notice and some credibility from the science side. The problem, however, is reciprocity - and whether the superficial hermeticism that has come to characterize design discourse is able to meet this challenge. No one, however, can deny that this is an interesting catalytic moment. While the aesthete side of design culture is attempting to hold its own as an offshoot of the fashion industry, the cracks in this position are widening. This is especially true at the urban scale - for example because urban marketing strategies are more and more interdependent with design strategies. And there is an evolving new academic amalgam involving business, science, and design, deliberately blurred in response to new problem sets and scientific boundaries.

The momentum for this approach, at least in the United States, is quickly strengthening around the global warming phenomenon. But our centers of learning will have to wake up to realities beyond the attractions of new business models. And in spite of new visualization and fabrication tools, designers will have to acknowledge

---

the city, we tend to leave the global blocs of the North and South, and China's and the U.K. emissions, as unquestioned. Some of the charities in the U.K. have been trying to question these emissions boundaries. A study by Christian Aid* has shown that if you isolate the activities of the U.K.'s top 100 companies, you get a very different emissions profile from the norm. The U.K. goes from generating 3.5 % of global emissions to about 15%. So math is where we can draw the boundaries. I wonder whether we could re-imagine a political map showing cities and their individual contribution to climate change.

* "Coming clean: revealing the UK's true carbon footprint, 2007"; Christian Aid, Policy Report, Climate Change http://www.christianaid.org.uk/resources/policy/climate_change.aspx

the reality that design culture seems to have regressed from where it was in the '60s - on the applications side of things, at least in terms of exploring a sustainable new "world model." Design culture needs to trace the same ground as the natural sciences have already covered during the last four decades - and more.

1. D.H. Meadows et al., *The Limit to Growth: A Report from Rome Club's Project on the Predicament of Mankind* (New York: Universe Books, 1972).

2. E.J. Mishan, *The Cost of Economic Growth* (New York: Praeger Publishers, 1967). Appendix C.

3. President George W. Bush and Prime Minister Tony Blair, remarks at The Cross Hall, November 20, 2003. Transcript, The White House, Office of the Press Secretary, http://www.whitehouse.gov/news/releases/2003/01/20030131-23.html

4. Kraft, J.C., et al.: "Paleographic Reconstruction of Coastal Aegean Archeology Sites," *Science*, 195, 1977.

5. Within the field of archeology, there is very interesting new work correlating built form and climatological science. For example, see Eberhard Zangger, *The Future of the Past. Archeology in the 21st Century* (London: Phoenix, 2002).

6. McKibben, B., *The End of Nature* (New York: Random House, 1989).

7. The restoration economy concept is well-summarized in: Storm Cunningham. *The Restoration Economy. The Greatest New Growth Frontier* (San Francisco: Berrett-Koehler Publishers, Inc., 2002) The economics of the new eco-enterprise is well analyzed in Herman E. Daley and Joshua Farley, *Ecological Economics. Principles and Application* (Washington DC: Island Press, 2004).

8. On the historical shift in environmental capitalism see also: Leo Paniteh and Colin Leys, *Coming to Terms with Nature. Socialist Register 2007* (London: The Merlin Press, 2006) Of particular interest is the essay by Neil Smith on "Nature Accumulation Strategy," 16-36.

9. On "canonical assumptions see Daley and Farley, *Op Cit*, pp. xxi-xxii.

10. Ulrich Beck and Johannes Willms, *Conversations With Ulrich Beck* Cambridge (England: Polity Press, 2004) 39, 183.

11. Serge Chermayeff first defined the concept of "design as catalyst" in reference to urban and environmental design. See Richard Plunz (ed.), *Design and the Public Good. Selected Writings, 1930-1980 by Serge Chermayeff* (Cambridge Mass: The MIT Press, 1982) 289-295.

CHAPTER 2

The Question of Environmental Justice
Julie Sze

I am examining the problems of urban climate change from an environmental justice perspective. In defining an environmental justice perspective on climate change, I highlight key examples where environmental justice and climate change are at the crossroads of recent social movement organizing, legislation, and community-based action. I identify both the necessity in bringing disenfranchised communities into the climate change conversation, and the roadblocks to doing so. Drawing on the work of political theorist David Schlosberg, I argue that to have communities as substantive and enthusiastic partners in the arena of climate change, scientists, policy makers and urban planners need to keep complex principles of justice at the core of their inter-sectorial initiatives, both distributive and participatory. As well, they must recognize diversity and difference in a myriad of forms: racial, nation-state, class and gender. In other words, urban climate action that takes an environmental justice perspective must acknowledge issues of process and justice. I begin by posing a few key questions. What is environmental justice? What is an environmental justice perspective on climate change? What are examples of an environmental justice perspective related to climate change in the United States? What are the potentials and the pitfalls regarding community-based science and planning as the world faces these changes?

Environmental Justice Defined

What is environmental justice? How is it defined? Nationally, the United States-based Environmental Justice Movement (EJM) emerged in the 1980s as a result of the confluence of events and

---

Bettina Menne
Climate change is a crucial aspect of my studies. But if you ask me as a citizen, maybe I would turn it another way around. I probably would put adaptation to climate change as most important for the European city today. But most likely I would first address social equity, and the question of our children. We have growing risks to our children and adolescents of transport - you know that's a big problem in Italy; the growing economic crises for households and job insecurity hamper adaptation. I would prioritize all these issues.

I probably would put adaptation to climate change as most important for the European city today

Domenico Cecchini
In the last ten to twelve years, Italy and Europe have seen an extraordinarily intense cycle of development and in-

reports that brought together the terms "environmental racism" and "environmental justice" into the public sphere and into policy discourses. Subsequent reports documented the "unequal protection" from environmental pollution by local, state, and national regulatory agencies. Environmental racism describes the disproportionate effects of environmental pollution on racial minorities, while environmental justice is the name of the social movement that emerged in response to this problem. Because it describes the disproportionate balance between high levels of pollution exposure for people of color and the low level of environmental benefits they enjoy, environmental racism can be defined as the unequal distribution of environmental benefits and pollution burdens based on race. The Principles of Environmental Justice were adopted at the 1991 First People of Color Environmental Leadership Summit and widely circulated. The principles were aimed at combating not only the abuses of corporate polluters, but also neglect by regulatory agencies, and class and racial biases in mainstream environmental groups. The Environmental Justice Movement gained mainstream political momentum when President Clinton signed an Executive Order on Environmental Justice in 1994, which mandated that all federal agencies generate agency-specific strategies to address the disproportionate pollution experienced by minority communities.

Issues engaged by the Environmental Justice Movement and defined as exemplifying environmental racism are broad and diverse; pollution impacts range across a broad spectrum of racial and ethnic groups in diverse geographic and social settings, from rural to urban. In the United States, these include: toxic/chemical pollution (oil refineries and petrochemical facilities; military pollution on Native lands); occupational exposures (in the garment industry and in Silicon Valley industry staffed by Asian and Latino immigrant women workers, including farm workers); health effects of poor

---

crease in building, and there has been a stunning creation of wealth and urban property revenues, in both our cities and elsewhere. Have these revenues found their way into other economic investments, as they did in the '70s and '80s; are they used to improve urban quality and promote sustainability?

Very little has been used for urban quality and sustainability. The assessments of the administrators of the whole structure of city government on the creation of urban wealth and its distribution should be transparent. It should be part of everyday administration, and no longer glossed over, or a matter for the big "developer," the important entrepreneur, or the mayor.

The assessments of the administrators of the whole structure of city government on the creation of urban wealth and its distribution should be transparent

housing such as lead poisoning; and consumption of contaminated fish by poor and immigrant communities.

Ideologically, the environmental justice movement seeks to expand the concept of "environment" to include public and human health concerns, in addition to natural resources such as air, land, and water. Social movement scholars have identified the key features of the environmental justice paradigm. The diverse issues, constituencies, and geography found under the umbrella "environmental justice" are linked through a world view or "environmental justice paradigm" that emphasizes an injustice frame. Pulido and Peña argue that the distinction between mainstream and environmental justice issues is based not only in issue identification, but on positionality, or a person's location within the larger social formation shaped by factors such as race, class, gender, and sexuality.[1]

Going Global

While environmental justice as a social movement and as a scholarly field arose originally in the United States, recent academic literature has begun to expand its scope to include other national contexts and the realm of the global. Schlosberg and Newell seek to revise the concept of environmental justice within a context of global environmental politics. By arguing that "recognition" of diverse cultural identities in a critical pluralism is a pre-condition for entry into the distributional system, Schlosberg allows environmental justice to encompass struggles for indigenous and popular movement rights, knowledge and identity around the world. Likewise, Schlosberg holds that both anti-globalization and people-centered development advocacy directed at multi-lateral institutions such as the World Trade Organization and the World Bank can also be seen as mobilizations against an encroaching "global monoculture." Schlosberg synthesizes his global view of environmental justice by

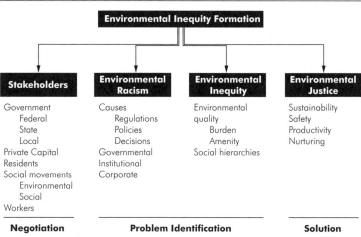

Fig 2.1. Paradigm for Environmental Inequity

Environmental Inequity Formation			
Stakeholders	**Environmental Racism**	**Environmental Inequity**	**Environmental Justice**
Government	Causes	Environmental	Sustainability
Federal	Regulations	quality	Safety
State	Policies	Burden	Productivity
Local	Decisions	Amenity	Nurturing
Private Capital	Governmental	Social hierarchies	
Residents	Institutional		
Social movements	Corporate		
Environmental			
Social			
Workers			
Negotiation	**Problem Identification**		**Solution**

Courtesy of Cynthia Rosenzweig

asserting, "It is not simply that the justice of environmental justice in political practice includes equity, recognition and participation: the broader argument here is that the movement represents an integration of these various claims into a broad call for justice."[2] Similarly, Newell argues that a global understanding of environmental justice must focus on "a broader set of questions than the role of institutions in global society and the interactions of state and non-state actors in global stage." Instead, Newell observes that when we "focus on intra and transnational social and economic divisions, looking for example at 'Souths in the North and Norths in the South', we have an entry point for assessing the importance of race and class to inequality in global environmental politics."[3]

Climate change is a similar field through which to analyze the dynamics of global environmental problems and policies with which to address them. As we know from the United Nations report, "Fighting Climate Change: Human Solidarity in a Divided World," the roots of climate problems and their impacts are globally differentiated and internally differentiated within the Global North.[4] For example, the 19 million people living in the New York region have a deeper carbon footprint than the 766 million people living in the 50 least-developed countries. Wealth and the production of wealth contributes disproportionately to climate change. Rich countries like the United States contribute many times the greenhouse gases per capita as poor countries, even as China's total carbon emissions exceeded the United States in 2007.

Internally within the Global North, similar power differentials are at play. Richer people in the United States use more resources than poor people. Very poor people account for half of the transit ridership in the United States and African Americans, Latinos, and Native Americans account for 60%. The Environmental Justice and Cli-

---

Matthew Nisbet

Though here in Europe it might seem strange to believe that addressing climate change can be thought of as a religious and moral duty, 30 to 40% of Americans are evangelicals, and other Americans are equally religious, and they talk about climate change in religious terms. In this context, using religious opinion leaders could be a very effective strategy for the climate change issue. But also in the United States and worldwide you cannot escape putting climate change in the mental box of economic development and economic opportunity, by focusing on development of new technologies, the potential for a carbon rate system, and even the economic benefits of a carbon tax.

Using religious opinion leaders could be a very effective strategy for the climate change issue

mate Change Coalition has made this point, and in 2002, produced a fact sheet that predicted Hurricane Katrina's social and environmental effects. It stated: "People of color are concentrated in urban centers in the South, coastal regions, and areas with substandard air quality. New Orleans, which is 62 percent African-American and two feet below sea level, exemplifies the severe and disproportionate impacts of climate change in the U.S." The term "transportation racism," for example, is given by sociologists including Robert Bullard, to the application of an environmental justice framework to transportation planning and policy. And given that transportation is a key component of disaster planning and evacuation, the fact that the stranded were poor, black, disproportionately elderly, young, and old, and without a private transportation alternative, reflects dimensions of environmental racism.[5]

What is Climate Justice?
Climate justice is a grassroots international social movement to implement solutions to global climate and energy problems using broad principles of justice as its foundation. The 2002 Bali Principle of Climate Justice (developed explicitly with the 1991 Principles of Environmental Justice as a model), outlines 27 core principles of what constitutes Climate Justice. While several principles discuss the global problem of climate change impacts, the core political and philosophical focus of this document is on the "most vulnerable" people and places.[6]

What does this focus on the most vulnerable mean in practice? What are the potentials and the pitfalls when it comes to community-based science and planning? What would it mean to really make communities - especially politically disenfranchised populations - partners in the face of climate change? It would mean centering their world view and taking the perspective of cultural recognition

---

Bruna De Marchi
And what about the language of ethics? The language of ethics is maybe not so widespread. It is mainly used as rhetoric: "we should be good" . . . "the new generation," and blah blah. Let us think of vulnerability. It is like a chain. In a chain, even if only one link of the chain is weak, then the whole chain is weak. Maybe we should be talking the language of common interest, and realizing that if a single part of us is vulnerable, we are all vulnerable.

Harriett Bulkeley
I am struck by the challenge of climate change being about how we deal with justice between communities in very different time scales and spatial scales.

at the core of climate change policy. In other words, to answer an environmental justice perspective on urban climate change, we must take seriously Lieven de Cauter's question, "What about Lagos?" As the world is increasingly urbanized in the Global South, and the gap between Global North and South is widening, we must start from the place of the most powerless, not from the perspective of the most privileged, in determining urban climate change policy. Rather than taking a technocratic view of disaster planning that could not imagine the world view and life experiences of a poor New Orleans resident who didn't own a car, climate policy must originate from the perspective of the disenfranchised.

As a final thought, as the world faces and prepares for the climate crisis, we must also refuse to engage in urban climate action and public policy that recreates social injustice. One example would avoid the fetish-ization of carbon neutral eco-cities and projects like Dongtan, near Shanghai, and take a skeptical approach to "green" initiatives that recreate numerous problems for the local population and depend crucially on political authoritarian regimes.[7]

1. L. Pulido and D.G. Peña, D.G., "Environmentalism and Positionality: The Early Pesticide Campaign of the United Farm Workers' Organizing Committee, 1965-1971," *Race, Gender & Class* 6 (1998) 33-50.
2. D. Schlosberg, "Reconceiving Environmental Justice: Global Movements and Political Theories," *Environmental Politics* 13 (2004) 517-540.
3. P. Newell, "Race, Class and the Global Politics of Environmental Inequality," *Global Environmental Politics* 5 (2005) 70-94.
4. United Nations, "Fighting Climate Change: Human Solidarity in a Divided World," *Human Development Report 2007-2008*, http://hdr.undp.org/en/media/hdr_20072008_en_complete.pdf (accessed July 2008).
5. J. Sze, "Toxic Soup Redux: Why Environmental Racism and Environmental Justice Matter after Katrina," *Understanding Katrina: Perspectives from the Social Sciences*, Social Sciences Research Council, 2005, http://understandingkatrina.ssrc.org/Sze/ (accessed July 2008).
6. http://www.ejnet.org/ej/bali.pdf (accessed August 22, 2008).
7. Dongtan is planned for Chongming Island in the Yangtze River near Shanghai. It has been widely publicized as a self-sufficient "green city." For an overview from the project builder see: http://www.arup.com/eastasia/project.cfm?pageid=7047

CHAPTER 3

Five Health Concerns
Bettina Menne

Introduction

There is now a strong global scientific consensus that the climate is changing and that if current trends of global warming continue, rising temperatures and sea levels and more frequent extreme weather events (heat -waves, storms, floods, droughts, cyclones, etc.) could lead to severe shortages of food and water, loss of shelter and livelihoods, and extinction of plant and animal species.[1]

Projected trends in climate -change-related exposures of importance to human health in urban environments are likely to:

- increase heat -wave related health impacts;
- increase flood -related health impacts;
- change food-borne disease patterns;
- change the distribution of infectious diseases;
- increase the burden of waterborne diseases, in populations where water, sanitation, and personal hygiene standards are already low, and
- increase the frequency of respiratory diseases due to higher concentrations of ground-level ozone in urban areas and changes in pollen distribution related to climate change.

These health effects will be unevenly experienced between and within different countries in the WHO European Region. Whether and how they will be experienced will depend on the adaptive capacity and actions of health systems, and local governments and the access different populations have to these services. Some of the measures might be efficient enough under current climates but might need to be strengthened or revised under much stronger or

---

Cynthia Rosenzweig
As a scientist, I feel a renewed sense of urgency, despite uncertainties. We need to streamline the delivery of the key information needs for urban decision makers. As scientists we start with questions first and then work to answer those questions. Now it is time to consider devising a new way of doing a rapid assessment in cities.

We need to streamline the delivery of the key information needs for urban decision makers

Bettina Menne
If you look at energy consumption and infant mortality, you see that above a certain level of consumption you don't improve infant the mortality rate anymore. Over a certain level of energy consumption, you don't increase life expectancy anymore. Conclusions about this are very arbitrary to make, but I think there is a message to be

accelerated climate change.

Cities have become areas of primary concern when it comes to the health effects of climate change; but they are also primary drivers of adaptation and mitigation efforts. Many cities in the eastern part of the WHO European Region are still recovering from the devastating effects of economic transitions on their budgets. They have had to deal with a complex array of linked economic, social, environmental, and health problems. Their difficult situations have forced them to choose between immediate action, unsustainable repair measures, and long-term exertions. Some huge residential buildings, for example, remain in poor condition due to a lack of maintenance and to inadequate construction standards. Contemporarily in Western Europe, there is a move out of the large cities into suburbs and smaller urban centers.

Action to adapt to climate change and mitigate climate change is essential in reducing vulnerability today and tomorrow. In order to protect population health when developing policies, measures, and strategies, it would be important to consider:
• Co-benefits: policies and action that reduce greenhouse gas emissions can also substantially reduce obesity, diabetes, heart disease, cancer, traffic deaths and injuries, and air pollution, e.g. through increasing physical activity
• Social and developmental goals: reducing energy consumption from food provides a link to five of the seven most important risk factors for premature death in the EU (blood pressure, cholesterol, Body Mass Index, inadequate fruit and vegetable intake, physical inactivity).
• Coherence: reducing air pollution offers probably the next largest health benefits, but not all greenhouse gas reduction measures benefit health.

---

taken home. Another issue is that we have to look at co-benefits. I don't like this distinction between "adaptation" and "mitigation." I know that for the conversation on climate change it is needed, but at the end of the day, if we look at health it doesn't matter so much. We have to look at the co-benefits to human health from the actions we take. But if we look at the examples again from the United Kingdom, the obesity epidemic has increased substantially. The use of cars has increased and the miles you walk on average have decreased. The distance walked per person per annum fell by 110 km during the last 30 years in UK, which is the equivalent of 1 kg of fat gain per year. I think this is quite an interesting equation. Even worse, if we look at the travel behavior, an average person makes three trips per day, spending

We have to look at the co-benefits to human health from the actions we take

Fig 3.1. Changes in temperature, sea level, N. hemisphere snow cover

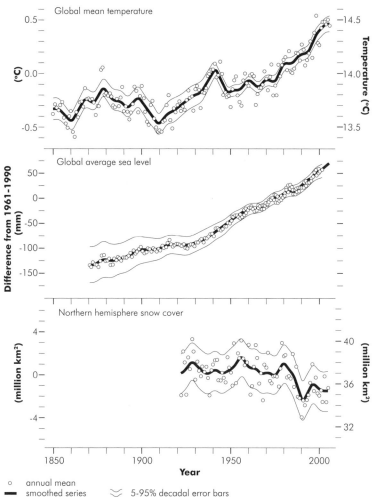

Source: IPCC Fourth Assesment Report, Working Group I Report The Physical Science Basis, 2007

Fig 3.2. Climate Change Economic Burden

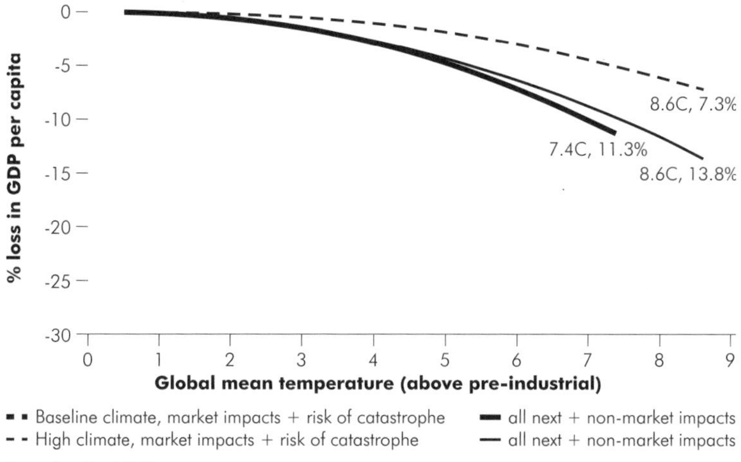

8.6C, 7.3%

7.4C, 11.3%

8.6C, 13.8%

Source: Stern Report, 2007

• Anticipatory action: increasing vigilance, early warning, early disease detection, and a number of primary and secondary preventive actions are needed to adapt to climatic changes.

Following are several examples with a major focus on mitigation.

1. Integrated sustainable transport systems

The transport sector is the second largest source of greenhouse gas (GHG) emissions in the European Region, after energy. Its relative share of emissions is increasing, particularly due to road transport and aviation. In addition to emitting GHGs, transport activities are also directly associated with a broad range of adverse health conditions. Policy choices in transport need to ensure health benefits. Emissions of GHGs are increasing as demand for passenger and freight transport offsets improvements due to better technology and stricter regulations. In the EU, for example, GHG emissions from transport could be 50% higher by 2030 than they were in 2000. Meanwhile, motorized transport is associated with several major causes of health problems.

• Less cycling and walking result in declining levels of physical activity. Physical inactivity is estimated to be associated with about 600,000 deaths per year in the European Region (excluding obesity), but cycling and walking can support more physically active lifestyles.

• Every year, more than 100,000 people die and more than 2.5 million are injured as a result of road traffic accidents, the leading cause of death for children and young people aged 5–24. While deaths and severe injuries are declining in the western part of the Region, they are still very high or even increasing in the eastern countries.

• Transport-related air pollution has been associated with several adverse health outcomes, including mortality, non-allergic respi-

---

one hour traveling. Only one in five trips is work-related. Five out of six trips begin and end at home, and 10% are not more than 1 km distance. Most of the trips are within 5 km, which is still within walking or cycling distance. So we have to look at the co-benefits in this sense. By increasing walking and cycling we are reducing the obesity epidemic, increasing traffic safety, and improving social behavior because we encourage people to be together, rather than having children sit by themselves at computers in their houses. I would like to ask you, to think about the co-benefits in more areas.

Julie Sze
I would like to emphasize the importance of public health. My own research is on asthma and public health activ-

By increasing walking and cycling we are reducing the obesity epidemic, increasing traffic safety, and improving social behavior

Fig 3.3. How Climate Change Affects Health

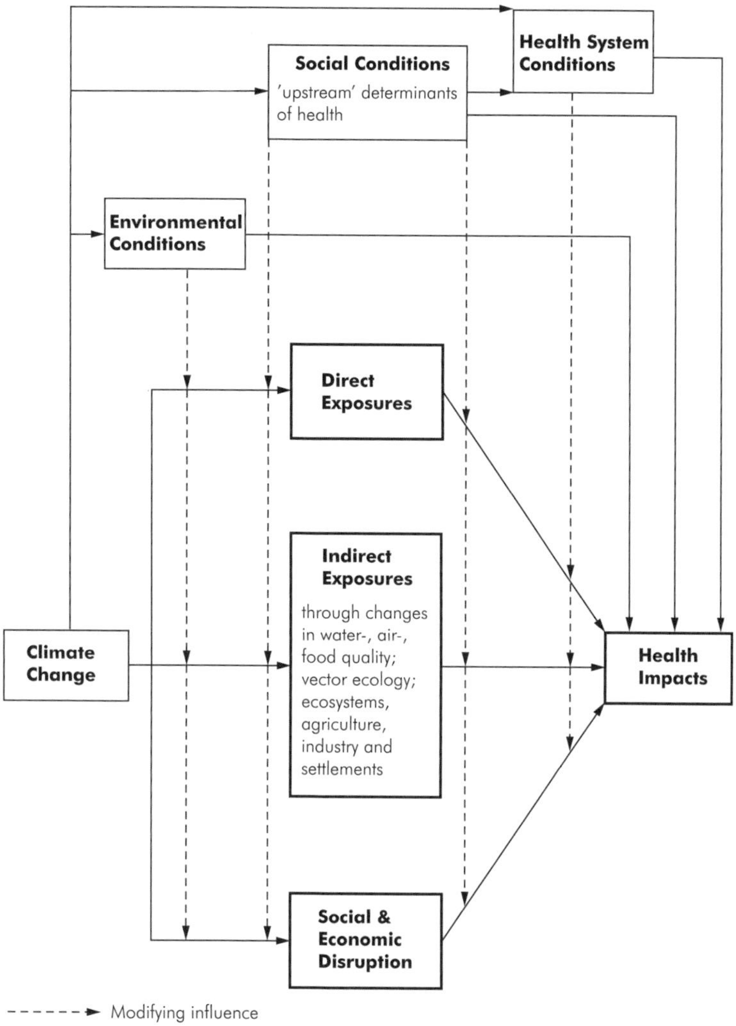

Health System Conditions
Social Conditions
'upstream' determinants of health
Environmental Conditions
Direct Exposures
Indirect Exposures
through changes in water-, air-, food quality; vector ecology; ecosystems, agriculture, industry and settlements
Climate Change
Health Impacts
Social & Economic Disruption
Modifying influence

Source: *IPCC Fourth Assesment Report, Working Group III Report Mitigation of Climate Change*, 2007

Fig 3.4. Transportation , Climate Change, and Public Health

Physical Inactivity
Noise
Air Pollution
Climate Change
Crashes
Psychosocial Effects

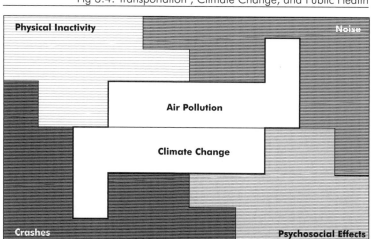

Courtesy of Bettina Menne

ratory diseases, cardiovascular diseases, cancer, and birth defects. Transport contributes significantly to the emissions of a large number of air pollutants, including precursors of ozone, PM, benzene, and, in some countries, lead.

• Long-term noise exposure is associated with a number of effects involving health and well-being. They include not only community reactions such as annoyance and disturbed sleep, but also physiological effects on, for instance, the cardiovascular system. Road transport is the most important source of noise pollution in urban areas.[2]

The burden of disease and death associated with transport is unevenly distributed across the Region, with low- and medium-income countries in the eastern and southern parts being more severely affected than high-income countries in the western part. Within countries, people from low socio-economic groups bear a disproportionate burden. Society largely bears these consequences, along with other such adverse effects as congestion, land use and environmental degradation, and together they are estimated to cost approximately several tens of billions of euros per year.[3]

Much of the policy response to date has focused on reducing transport-related emissions of GHGs through technological improvements, mostly addressing fuel type and quality and engine efficiency, with the support of public incentives. These policies, however, have demonstrated a limited capacity to control the continuing increase in transport-related GHG emissions. This limitation is partly due to the fact that the increase in kilometres travelled by both passengers and freight rapidly outweighs improvements in vehicles and fuel efficiency. Moreover, despite the economic incentives provided by many governments, cleaner technologies require several years of continued investment before any measurable results appear. For

---

ism, but one of the largest problems is that we live in a country that has no national health care.

Bettina Menne
Don't you overstate the power of the public health question? In October 2007, Julie Gerberding, Director of the Center for Disease Control, gave her testimony at the U.S. Senate on the public health effect of climate change. What gained the attention was not that she spoke about climate change effects on human health. In fact, that was subsequently erased from the talk. What gained public attention was that the White House administration canceled seven out of twelve pages of the hearing. Therefore, it was not the health factor that captured the attention, but it was the fact, that changes to the text were done by the

One of the largest problems is that we live in a country that has no national health care

these reasons, strategies that seek to manage transport demand – for example, by reducing the need to travel, promoting public transport, or discouraging the use of private cars through congestion charges or parking fees – can be very attractive. In fact, many such policies offer the possibility of simultaneously addressing the problems of GHG emissions and the negative health-related effects of transport, thereby maximizing collateral health benefits. An additional advantage of these mobility-management policies is that they can be applied locally and require less lead time and investment for implementation than developing new infrastructure to accommodate a growing number of vehicles.

A wide variety of practical experience already exists to support the development of policies combining GHG reduction with other health benefits, such as congestion charges, safer cycling paths, and monetary incentives. These experiences need to be more systematically documented, reviewed, shared, and integrated into transport planning, particularly at the urban level, where such measures have the greatest potential.[4]

2. Spatial planning and the built environment
Spatial planning and the built environment have a central role to play in both adaptation and mitigation to climate change, and they impact both directly and indirectly on health. Residential and commercial buildings contribute to GHG emissions. People pass many hours a day in buildings. Buildings can contribute to ill-health. Measures to reduce GHGs need to aim also at improving health.

In 2004, direct emissions from the housing/building sector (excluding emissions from electricity use) were one of the major sources of GHG emissions, accounting for about 7% of the global total. Emissions from this sector are projected to rise significantly in the

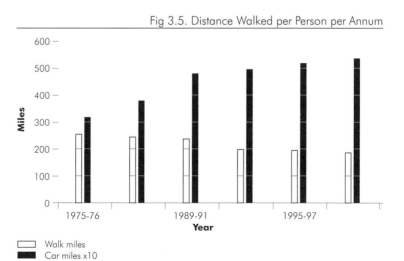

Fig 3.5. Distance Walked per Person per Annum

Walk miles
Car miles x10

Source: Fox and Hillsdon presentation to goverment Foresight policy development programme on obesity

near future. Residential energy use can also affect health through its influence on indoor temperatures and air quality.[5] In 2007 the International Panel on Climate Change (IPCC) estimated that the building sector has the greatest possibility for reducing GHG emissions at a low cost or even a net savings. Energy efficiency and thermal performance are key aspects of health protection in the built environment.

Energy-saving measures suggested by the IPCC, include efficient lighting, thermal insulation, the use of solar energy for space and water heating, and improved cooking stoves where needed. They all present opportunities to reduce energy consumption and associated GHG emissions, but they generally require active government policies to make them a reality.[6] Public information campaigns to affect people's behaviour with respect to appropriate heating, cooling, and ventilation should remain a priority issue regardless of infrastructure improvements.

The long lifetime of buildings means not only that their designs have implications for associated GHG emissions many decades into the future but also that they need to be able to perform all desired functions (such as natural cooling) in a climate that may change to include greater extremes of heat, storms, and precipitation. One special concern relates to indoor temperatures and air quality in health care facilities and nursing homes. Health-system action will be needed to ensure adequate planning of locations for health care and nursing institutions, as well as for the thermal and indoor air protection of their facilities.

3. A shift to renewable energy sources
Social systems, economic development, and the populations' health status are strongly dependent on electric power generation and dis-

---

government.

Harriett Bulkeley
I think that the cumulative effect of lots of small pieces of regulation are important, and also the way in which we think about our urban infrastructures. It is a kind of hidden hand of the state and companies to create a climate-changed city which we don't even really notice as being changed.

The cumulative effect of lots of small pieces of regulation are important, and also the way in which we think about our urban infrastructures

Richenda Connell
We need a combination of change management and risk management approaches. The latter is about how to make good and robust decisions on adapting to climate change, and the former is about how you change

tribution. However, electric power generation and distribution is associated with a wide range of direct and indirect adverse health effects associated with the whole fuel cycle. Electric-power production is associated with various health consequences for the industry workforce and the general population. These effects may arise during fuel extraction (mining or drilling); power generation, transmission, distribution, and consumption; or waste-product disposal. Some health effects can occur within hours or days (e.g. the acute effects of air pollution), while others may take decades or centuries (e.g. those associated with global climate change or radiation). For example, in monetary terms, the external costs from electricity generation in the 15 EU countries in 1998 were equivalent to 1% of the total GDP of the EU. Of this total, about 90% was caused by health effects. The biggest health effects arise from coal, oil, and gas power cycles. By comparison, renewable sources such as wind and solar energy have limited adverse health effects. The production of electricity from nuclear sources remains a concern for health, because of the risks from nuclear accidents, including those potentially caused by security accidents, and nuclear waste disposal.

Today, there is no doubt that energy consumption needs to be reduced. Health can contribute in driving the agenda. It can make health and climate-friendly choices easier with regulatory measures that can incorporate health impact assessments of energy choices, links to air quality guidelines and mandatory energy labelling of cars, homes, appliances, and other products. In view of the great dependency on energy imports of most European countries, the assurance of energy security is a universal policy objective. From a health perspective, residential access to clean, affordable, and reliable energy is fundamental to promote and protect the populations' health status, consistent with the aims of the United Nations' Mil-

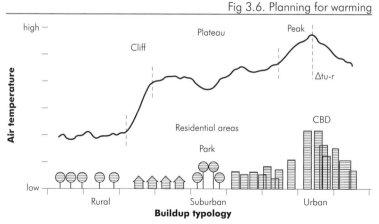

Fig 3.6. Planning for warming

Δtu-r = Temperature difference between urban center and rural area
CBD = Central Business District

Source: Bitan, 2003

lennium Development Goals.

In the middle of plenty, still some populations in Europe do not have access to commercial energy. The indoor use of solid fossil fuels such as coal or wood for heating presents a major health threat in at least 14 of the 53 countries in the European Region, and some population groups face these problems in at least 6 additional countries. Solid fossil fuels contribute substantially to the increase of GHGs, and they accounted for about 13 000 deaths among children aged 0–4 in 2004.[7]

4. Greening health services
The health sector has a challenge – and an opportunity – to demonstrate its leadership and moral authority in dealing with climate change by acting to reduce its own carbon footprint. A health system can demonstrate good corporate practice and citizenship while utilizing its size and financial power to find and demonstrate innovative ways to reduce its environmental footprint, improve health, and save money.

Several areas have been identified for actions that can benefit the social, environmental, and economic conditions within which the health system functions: energy, transport, procurement (of food, etc.), buildings and landscape, employment and skills, and community engagement. Community engagement can help to inform and influence action and communities can lead examples of good practice.

5. The power of local action
In a single day, a European city of one million inhabitants uses an estimated 320,000 tonnes of water, 11,500 tonnes of fossil fuels and 2,000 tonnes of food. It also produces 300,000 tonnes of

---

institutions and people so that they take action, and feel empowered to do so. You need the relevant government decision makers. You need the relevant businesses. You need them to be advised by the best researchers and consultants. You need to bring these groups together.

Harriett Bulkeley
How do you invest addressing climate change into the fabric of what the city normally does? This question involves leaders taking on some unpopular decisions. When you buy a flat in London you don't much care how it is heated as long as it is heated. So, the Mayor has said to all the developers "You have got to put in combined heat and power." That is the cost the developer has to pay, and you are serving the public interest.

You need the relevant government decision makers. You need the relevant businesses. You need them to be advised by the best researchers and consultants

wastewater, 25,000 tonnes of CO_2 and 1,600 tonnes of solid waste. Each year, an estimated 70–80% of European Region cities with more than 500,000 inhabitants exceed WHO air quality standards at least once, and about 90% of the Region's urban population is exposed to potentially health-damaging effects from air pollution.[8] Some of the first negative effects of heat -waves associated with climate change appeared in cities, including the deaths of more than 70,000 extra people in European cities in summer 2003. The heat-island effect may contribute to the greater vulnerability of city dwellers.

Policy options include early warning systems, health system preparedness and response, urban and community planning, and housing improvements. A comprehensive adaptation plan should involve multiple public entities, such as city management, the public health department, social services agencies, emergency medical services (or their rural equivalents), and civil society. Communications should be developed to advise people of appropriate behaviours. City planning should ensure that jobs, shops, and services are close to residential areas. The development and maintenance of green spaces is also of fundamental importance.

A baseline health and emissions inventory can be developed in conjunction with a local action plan. Such a plan might include energy-efficiency improvements to municipal buildings and water-treatment facilities, streetlight retrofits, public transit improvements, renewable energy installations, and CH_4 recovery from waste management. Long-term climate forecasts should be taken into account in constructing new buildings and planning new neighbourhoods, to provide as much thermal comfort and protection from extreme weather events as possible. An important component of new construction should be employing the best possible methods and ma-

---

Cynthia Rosenzweig
We have to engage all the actors from the beginning. We have to engage with communities; utilizing the techniques of conversation and discussion, but always through respectful communication. This involves interactions among the different groups in city societies, so that the communities, the public sector, the private sector, and the artists are all engaged. We realize that there is going to be creative tension, but that will help to lead us to truly sustainable solutions.

terials for space cooling. Relying on energy-intensive technologies such as air conditioning is not sustainable and can be considered maladaptive. Finally, it is important to monitor progress and report results, for example by installing roadside pollution meters and announcing the readings to the public on a daily basis.

Local communities and cities are in an optimal position to establish links across sectors and departments. Addressing the health impacts of climate change provides an opportunity for the integration of public health and climate change knowledge. Integration requires reciprocal understanding of terminology, goals, and methods. Beyond this, it requires working together to achieve the goal of reducing deaths, disease, and disabilities. City and local community leaders can exercise their powers. As policy-makers, the power can be used to institutionalize the changes discussed above. For example, one can insist that all new housing and all transport meets certain standards that protect health and the environment. Or, one could introduce traffic-reducing measures such as congestion charges (toll fees for entering central city areas), bicycle lanes and park-and-ride to limit CO_2 emissions. Today there are already a number of leading examples of cities and local community action in reducing GHG emissions and adapting to climate change while improving health. Sharing lessons learned from these experiences could contribute to collective action world-wide.

This paper was prepared in collaboration with Francesca Racioppi, and Pier Paolo Mudu.

1. IPCC, 2007: Climate Change 2007: *Synthesis Report. Contribution of Working Groups I, II and III to the Fourth Assessment Report of the Intergovernmental Panel on Climate Change.*Core Writing Team, Pachauri, R.K and Reisinger, A., eds., IPCC, Geneva, Switzerland.

2. F. Racioppi et al., "Preventing road traffic injury: a public health perspective for Europe," WHO Regional Office for Europe, Copenhagen, 2004, http://www.euro.who.int/document/E82659.pdf (accessed March 14, 2008).

3. Pan-European Program, "Transport, Health and Environment Pan-European Programme," United Nations Economic Commission for Europe/WHO Regional Office for Europe, Geneva/Copenhagen, 2008, http://www.thepep.org (accessed March 14, 2008).

4. N. Cavill, S. Kahlmeier, and F. Racioppi, "Physical Activity and Health in Europe: Evidence for Action," WHO Regional Office for Europe, Copenhagen, 2006, http://www.euro.who.int/document/e89490.pdf (accessed March 14, 2008).

5. IPCC, "Summary for Policymakers," IPCC, 2007: *Climate Change 2007: Mitigation. Contribution of Working Group III to the Fourth Assessment Report of the Intergovernmental Panel on Climate Change,* Cambridge University Press, Cambridge, United Kingdom and New York, NY, USA.

6. M. Levine et al., "Residential and Commercial Buildings," IPCC, 2007: *Climate Change 2007: Mitigation. Contribution of Working Group III to the Fourth Assessment Report of the Intergovernmental Panel on Climate Change,* Cambridge University Press, Cambridge, United Kingdom and New York, NY, USA.

7. K.R. Smith, S. Mehta, and M. Maeusezahl-Fuez, "Indoor Air Pollution from Household Use of Solid Fuels," in M. Ezzati et al., eds. *Comparative Quantification of Health Risks: Global and Regional Burden of Disease Attributable to Selected Major Risk Factors,* (Geneva: World Health Organization, 2004)1436–1493.

8. European Commission, *WALCYING: How to Enhance WALking and CYcliNG Instead of Shorter Car Trips and to Make These Modes Safer – Final Project Report,* Office for Official Publications of the European Communities, Luxembourg, 1997. http://cordis.europa.eu/transport/src/walcyngrep.htm (accessed March 25, 2008).

CHAPTER 4

Cities and Governance
Harriett Bulkeley

Cities have emerged as key players in the governance of climate change. While conventional political analysis has tended to neglect the role of these local actors with respect to the global problem of climate change, the proliferation over the past decade of city and city-network initiatives to address the problem is increasingly hard to ignore. Reflecting this trend, a growing body of research has drawn attention to the ways in which cities have engaged with the issue of climate change.[1] This work has documented initiatives that are taking place at the city level and the challenges which have been encountered, primarily in relation to climate change mitigation and with a focus on cities in the "North." However, in the main, this research has taken the urban as a relatively unproblematic category, and has not engaged with the shifts in urban governance documented in the wider literature on urban studies or emerging debates concerning critical readings of urban metabolism.

This paper seeks to engage with these debates and the insights they provide into the governance of climate change mitigation and adaptation. First, the scope and nature of climate change governance at the city scale will be reviewed. Here, the focus will be on reviewing the evidence for urban climate governance, and on an analytical framework for assessing the modes through which climate change is being governed.[2] Second, the paper will argue for a need to embed our analysis in a fuller understanding of the social and material dynamics of cities. Here, the ways in which changes in the systems of energy provision are working for and against local climate change strategies will be highlighted. In conclusion, the paper will

---

Matthew Nisbet

Compared to Europe, the difference between our two political systems is that in the United States there is a tremendous partisan divide on the issue....Before policy makers overcome that divide and actually make policy action on climate change a priority, we have to see climate change starting to poll as a public priority, or at least hear from the constituents on that issue – and the politicians aren't seeing that right now.

Before policy makers overcome that divide and actually make policy action on climate change a priority, we have to see climate change starting to poll as a public priority

Richenda Connell

In the UK there is an organization called the "UK Climate Impacts Programme" (UKCIP*). It arose out of recognition that despite international and national studies that talked about the impacts of climate change, the issue wasn't

reflect on the implications for the interactions between processes of urbanization and climate governance.

Urban Climate Change Governance: Adaptation - Mitigation
The term governance has become almost ubiquitous in fields as diverse as urban studies and international relations. In its broadest interpretation, and the one adopted here, governance is simply used to "refer to the modes and practices of the mobilisation and organisation of collective action."[3] Governance is here a catch-all term, conceived as "the instituted process" that is created by and serves to guide processes of governing.[4] Cities, or more accurately their municipal governments and partners, have become a critical site for the governing of climate change over the past decade.

A recent survey conducted by the Tyndall Centre for Climate Change Research in the UK has found that currently, more than 2,000 municipal governments worldwide have made some form of commitment to address climate change action and that this dominates the activities of non-state actors in relation to climate change.[5] Activity is concentrated in OECD countries but there has also been growing engagement in developing countries. Although the actions being taken by cities are primarily declarations of intent, the same Tyndall survey found a core of approximately three hundred cities that were taking a more active stance, and that in these cities the commitment for emissions reductions will be in the order of 20% by 2012 through a mixture of policies and measures primarily related to the energy sector. Perhaps rather strangely, given the distance (in spatial and temporal terms) between the mitigation efforts of cities and their effects on the global atmosphere, it has been mitigation rather than adaptation that has to date dominated the urban climate change governance agenda.

---

tangible enough for organizations working at a regional or local level. UKCIP was quite a small organization and was not set up to engage directly with the general public. Instead, it worked mainly at the level of regions and local authorities. It worked with the local policy makers, local decision makers, and local water companies. We found that if we presented information to people about climate change that was at a national level, it didn't seem relevant to them. It was only when we brought tech impacts of climate change down to their local level – and we named places, towns, and landmarks that they identified with, that it had a totally different resonance with them. Then there was a turning point in the year 2000, when we had very widespread flooding across the UK. For some reason the newspapers chose to say "This is cli-

We found that if we presented information to people about climate change that was at a national level, it didn't seem relevant to them

Governance Challenges I: Vulnerability and Adaptation

Climate change represents one facet of the vulnerability of urban areas, particularly for the mega-cities of the Global South. In the fourth International Panel on Climate Change (IPCC) report, Wilbanks et. al. provide an excellent summary of the inter-linkages between climate change and urban vulnerability:

> Climate change is not the only stress on human settlements, but rather it coalesces with other stresses, such as scarcity of water or governance structures that are inadequate even in the absence of climate change ... These types of stress do not take the same form in every city and community, nor are they equally severe everywhere. Many of the places where people live across the world are under pressure from some combination of continuing growth, pervasive inequity, jurisdictional fragmentation, fiscal strains and aging infrastructure.[6]

However, urban vulnerability is not confined to the Global South. In its interpretation of the climate change challenge, the Stern Review suggests that "many of the world's major cities (22 of the top 50) are at risk of flooding from coastal surges, including Tokyo, Shanghai, Hong Kong, Mumbai, Calcutta, Karachi, Buenos Aires, St Petersburg, New York, Miami and London."[7] Both reports highlight the differential geographies of vulnerability, created by the combined differences in the physical environments of cities, processes of urbanization, socio-economic exclusion and institutional structures. In the main, however, vulnerability is regarded as an issue for cities in the Global South. Recent climate-related events (e.g. Hurricane Katrina and the European heat waves of 2003) have started to challenge this received wisdom. Nonetheless, this association with vulnerability and cities in the Global South has meant that adaptation has until very recently remained off the climate governance agendas of most cities in the developed world.

It was only when we brought the impacts of climate change down to their local level – and we named places, towns, and landmarks that they identified with, that it had a totally different resonance with them

mate change," – they blamed climate change for causing the floods, and climate change was all over the newspapers. When we had floods last year in Oxford, there was a demonstration in the city with people shouting "Gordon Brown, Gordon Brown, climate change is in our town." People really felt an immediacy about climate change – when it's something that's happening to them now.

*The UK Climate Impacts Programme (UKCIP) was established in 1997 to help coordinate scientific research into the impacts of climate change, and to help organizations adapt to those unavoidable impacts. The majority of UKCIP's funding is from the Department for Environment, Food and Rural Affairs. UKCIP is based at the Environmental Change Institute at Oxford University.

As the issue of governing adaptation gains political momentum, policy prescription has been rather simplistic and lacking an acknowledgment of the complexities of urbanization that will shape both vulnerabilities and the possibilities of adaptation. There is a lack of engagement with the politics of governing for adaptation, including the context of decision-making and the multiple actors and coalitions involved in contemporary urban governance. A distinct contribution that a new agenda on cities and climate change could offer is to grapple with the processes and politics of urban vulnerability and adaptation.

Governance Challenges II: Mitigation

Cities, it is suggested in the Stern Review, may "account for 78% of carbon emissions from human activities."[8] Despite the significance of urban emissions and the relatively long history, at least in some cities, of urban engagement with the mitigation agenda, this picture remains concerning engagement with the processes and politics of urbanization. Urban climate change policy tends to be conceived – like environmental policy more generally – as a policy sector in its own right, with little account taken of how other policies (for economic growth, say) or processes (of migration, for example) are shaping local emissions profiles and enabling/constraining the opportunity space for action. If we consider the main findings from over a decade of research into the governance of mitigation within cities, we can see that this approach to policy prescription is inadequate for three main reasons.

First, the "urban" governance of climate change is not bound by city limits. It is multilevel in the broadest sense – not only about interactions among multiple levels of government, but also what has been termed "Type II" multilevel governance or horizontal networking across scales, which creates new political arenas not neces-

---

David Burney
At the local government level we do see immediacy; in the issues of waste disposal, air quality, water supply, energy supply, air pollution, and more recently, flooding, arguably all connected to climate change. Now at the local government level you are seeing a coalition between mayors of individual cities like New York, Chicago, Seattle, around global warming. In the absence of action by the federal government, the State of California has now begun its own air quality and energy constraints on the car industry, resulting in America's favorite pastime, litigation, between the State and the Federal government.

At the local government level you are seeing is a coalition between mayors of individual cities like New York, Chicago, Seattle, around global warming

Marianella Sclavi
We are living a very big transformation, and we must

sarily tied to any particular territory. Second, while governing tends to focus on the municipal authority, there are a range of other players involved – national and regional governments, corporate actors, foundations, and "hybrid" networks. This is governance in a very real sense. While the term may be so over-exposed as to lead to its dismissal, this calls for further attention to just what it might mean to seek to address climate change – and potentially a very large source of emissions – in the absence of the powers traditionally associated with (municipal) government.

Third, these new governance arrangements do not employ one set of policy approaches to the issue of climate change. Rather, as research conducted by myself and Kristine Kern has found, there are multiple modes of governing at work (Table 4.1). Our research suggests that it is the "self-governing" and "enabling" modes that dominate the urban climate change governance landscape. In the first mode, the municipal government itself becomes subject to pressures to address climate change – as a good citizen, as a means of leading by example, and because of the economic efficiencies to be gained. It is this last discourse that has provided the strongest rationale for action within the municipal government arena.

In order to access alternative sources of emissions the municipality must, as indicated above, work with a range of different partners in an "enabling" mode. Two examples from London illustrate the point. First, a "green concierge" was piloted with 40 Lewisham homes in partnership between the Mayor, the London Development Agency, the Energy Saving Trust, and a private provider. This is a fee-based service and includes an energy audit (in varying levels of depth according to the available budget) as well as project management for the implementation of the measures the homeowner chooses.[9] The pilot is now being rolled out to homes across Lon-

---

The process of getting out of deeply established frames entails the ability to deal with absurdity

change completely our frame of mind. My favorite authors (Gregory Bateson, Kurt Lewin, Arthur Koestler) are scholars who thought a lot about the dynamics of change: the difference between change inside a frame and changing the frame itself. The process of getting out of deeply established frames entails the ability to deal with absurdity, because this escape can sound like a completely irrational, nonsensical thing to do. For Koestler creativity is nonsense with a pattern. Only when we have built a new frame do we see the process as "rational." There is an ocean of difference between "chosen nonsense" as the other face of creativity, and "suffered nonsense" as the result of too much obstinacy in linear thinking. In addressing climate change we must acknowledge that we are dealing with a situation that looks and sounds "im-

don.[10] While, as in most cities, there has been a focus on the domestic sector as a source of emissions, there is increasing interest in commercial spaces.

The C40 network, with financing from the Clinton Foundation, has launched a "Better Buildings" program that promises to bring together cities, building owners, banks and energy, service companies to make changes to existing buildings to reduce greenhouse gas emissions. This model will promote energy service companies in making investments in energy efficiency and gaining a share of any cost savings made.[11] Such examples demonstrate the multi-level, multi-actor and multi-modal approach to the urban governance of mitigation that is taking place. Key challenges in this field are partly methodological – how to trace and assess the roles of these web-like structures – and partly evaluative, in terms of understanding the impact on emissions such initiatives are having in real terms. At the same time, some relatively fundamental questions about how we understand the dynamics and whereabouts of power are raised. Although this is an issue for the analysis of climate governance more generally, the urban arena offers a promising lens through which to scrutinize this question.

Urban Material Worlds
There remain, then, serious conceptual and methodological challenges to our understanding of the urban governance of climate change in terms of relating to the broader sets of processes that are shaping contemporary urban places and in examining the politics of urban responses. A further set of challenges emerges when we take into account not only the economic, social, and political make-up of cities, but their physical infrastructures. By and large, addressing the adaptation and mitigation challenges of climate change will require some fundamental shifts in the infrastructures that we take for

---

possible," and also "inconceivable" for its catastrophic dimensions, and that for "normal people" (nearly all of us), what needs to be done to reverse it looks and sounds rightly absurd. If we admit this, it will already look more sensible. We understand certain kinds of messages only if we practice them: it is not only learning by doing but also understanding by doing. It is the kind of change that requires a special elaboration of emotions, and it must become common knowledge.

David Burney
To a large extent our central government is constrained by many special interests. In the United States, Exxon Corporation last week announced a record 41 billion dollar profit, the largest ever profit by an American corporation.

We understand certain kinds of messages only if we practice them: it is not only learning by doing but also understanding by doing

	Self-governing	Governing by authority	Governing by provision	Governing through enabling
Energy	Energy efficiency schemes within municipal buildings (e.g. schools) Use of CHP within municipal buildings Purchasing green energy Procurement of energy efficient appliances Eco-house demonstration projects Renewable energy demonstration projects	Strategic planning to enhance energy conservation Supplementary planning guidance on energy efficiency design Supplementary planning guidance on CHP installations of renewables Supplementary private contracts to guarantee connection to CHP or renewable energy installations (Germany)	Energy efficiency measures in council housing Energy Service Provider (Stadtwerke) (Germany) Energy Service Companies (UK) Community energy projects (UK)	Campaigns for energy efficiency Provision of advice on energy efficiency to businesses and citizens Provision of grants for energy efficiency measures Promote the use of renewable energy Loan schemes for PV technology HECA report (UK)
Transport	Green travel plans Mobility management for employees Green fleets	Reducing the need to travel through planning policies Pedestrianization Provision of infrastructure for alternative forms of transport Workplace levies and road-user charging (UK)	Public Transport Service Provider (Verkehrsce-trlebe) (Germany)	Education campaigns on alternatives Green Travel Plans Safer Routes to School Walking Buses Quality partnerships with public transport providers
Planning	High energy efficiency standards in new buildings Use of CHP and renewables in new council buildings Demonstration projects - house or neighborhood scale	Strategic planning to enhance energy conservation Supplementary planning guidance on energy efficiency design Supplementary planning guidance on CHP installations or renewables Supplementary (private) contracts to guarantee connection to CHP or renewable energy installations (Germany)		Guidance for architects and developers on energy efficiency Guidance for architects and developers on renewables
Waste	Clean up all contaminated land in New York City.	Provision of sites for recycling, composting, and "waste to energy" facilities Enable methane combustion from landfill sites	Recycling, composting, re-use schemes Service Provider (Stadtwerke) (Germany)	Campaigns for reducing, reusing, recycling waste Promote use of recycled products

Courtesy of Harriett Bulkeley

Table 4.1. Modes of Local Governance and Climate Change Policy

granted – those that supply energy, transportation, water, sewerage, and waste services. To illustrate this point, I focus here on the energy system and its role in climate change mitigation. An important starting point is that such systems are socio-technical; that is, they include technical components, such as the technologies that generate energy (e.g. solar panels on a roof), and social components, the institutions that guide the development of these technologies (e.g. planning laws) and the behaviours through which energy demand is produced (e.g. using hot water). Over the past couple of decades such systems have undergone some profound changes.[12]

Institutionally, there has in many, but not all, parts of the world been a decoupling of energy utilities from state authorities. Through various processes of liberalization and privatization, energy services are now provided by a range of different operators. At the same time, the nature of energy policy has shifted away from a concern purely with the reliability and cost of supply, toward issues of energy security and climate change. In establishing the regulatory contexts and policy goals within which it is bound, the state is far from absent from the maintenance and dynamics of the energy system. In this context, municipal climate change politics becomes enmeshed in the activities and decisions of a range of multi-national corporations and national governments. The increasing complexity and fragmentation involved, coupled with the strategic battles to be won in order to place climate protection on the agendas of a range of actors, can serve to undermine the potential for municipal action.[13]

There are however also opportunities. The first appears in terms of the social and institutional means through which energy services are provided. The emerging concentration on enabling modes of governing for climate protection creates particular capacity challenges

---

Exxon Corporation has a vested interest in the production of fossil fuel energy, and of course, resists all attempts to invest in alternative energy, seeking to persuade the central government to give it additional rights to explore in Alaska, to mine, and so on. Similarly, in Detroit the car industry has assiduously resisted constraints on fuel efficiency for cars, on development of alternative fuels and on development of electric cars. These special interests have very powerful control over parts of Congress because they fund the election campaigns of the politicians. So, however the debate might go, whatever the sympathies of the Democrats and the Republicans, they are very much constrained by these forces.

These special interests have very powerful control over parts of Congress because they fund the election campaigns of the politicians

for local government: to create financial incentives for action; to persuade others of the need for action; and to coordinate action across different arenas and sectors in order to generate new governing capacities. However, where this has been successfully managed, new approaches can emerge to energy supply, which are arguably more socially just and environmentally sensitive. This is the case in Melbourne, Australia, where a group of local authorities have used their collective purchasing power to provide access to green power to normal/low tariff constituents (the scheme is termed "Community Power"). Second, the splintering of the energy service grid is also creating technical opportunities as "innovative technologies of power generation, system control and energy end-use have become competitive."[14] These technological developments are reflected in the increasing interest in the potential of "micro-renewables" and decentralized energy systems – sometimes termed "embedded generation." For example, in the London borough of Merton, the "Merton rule," as it has become known, requires developers to ensure at least 10% of all energy production for new development comes from renewable energy equipment on site.[15]

Concluding Remarks

The governing of climate change at the urban scale is not taking place within a social, political, economic, or material vacuum. Other processes of urbanization are shaping the conditions of possibility for acting on climate change. To date, analysis of the urban politics of climate change has tended to overlook these rather profound transformations in what the city is and how it operates. The brief analysis presented here suggests that four points are particularly important as we move beyond the current "crossroads" towards future research and policy agendas.

First, in relation to both mitigation and adaptation, it is clear that

---

Domenico Cecchini
In the '50s and '60s, in the era of the Club of Rome, questions such as where and how energy revenues are used were on the agendas of technical, cultural, and political debate. They gave rise to a body of legislation that in time has become outdated. Do these revenues find their way into unrelated economic investments, as they did in the '70s and '80s, or are they used to improve urban quality and promote sustainability? Very, very little is used for urban quality and sustainability.

we are dealing with a situation of urban "governance" rather than "government." This will have significant implications for scientific efforts at the urban level, the stakeholders who should be involved, as well as the ways in which politics is conceived. Second, the potential for intervention to address climate change is shaped by a range of political and material dynamics that we barely grasp. We need to widen our horizons to understand the complex processes that are shaping urbanization in real time, as well as potential future trends. Third, and related, this involves an understanding and appreciation of the city not as some bound entity which we can locate on a map, but as a set of networked processes that come together in particular conjunctions in particular urban places. Methodologically this will mean a need to "follow the network" rather than be bounded by city limits. Finally, although we need to take account of the significant scholarship "out there" on the city, we also have a case to make that such approaches need to take seriously the urban governance of climate change, not least because it may be the "black swan" that refutes existing theory about the nature of urbanization.

1. For a summary of this engagement, see: Betsill, M.M., and Bulkeley, H., 2007: "Looking Back and Thinking Ahead: A Decade of Cities and Climate Change Research," *Local Environment: the International Journal of Justice and Sustainability*, 12 (5): 447-456.
2. Bulkeley, H. and Kern, K., 2006: "Local Government and Climate Change Governance in the UK and Germany," *Urban Studies*, 43 (12): 2237-2259.
3. Coafee, J. and Healy, P., 2003: "'My Voice, My Place': Tracking Transformations in Urban Governance," *Urban Studies*, 40 (10): 1979-1999.
4. Lowndes, V., 2001: "Rescuing Aunt Sally: Taking Institutional Theory Seriously in Urban Politics," *Urban Studies* 38 (11): 1953-1971.
5. Mann, P. and Liverman, D., 2007: An Empirical Study of Climate Mitigation Commitments and Achievements by Non-State Actors, Free University Amsterdam, May 2007.
6. IPCC, 2007: Climate Change 2007*: Impacts, Adaptation and Vulnerability. Contribution of Working Group II to the Fourth Assessment Report of the Intergovernmental Panel on Climate Change*, Cambridge University Press, Cambridge, UK.
7. Stern, N., 2006: *The Economics of Climate Change: The Stern Review*, HM Treasury and the Cabinet Office, London,
http://www.hmtreasury.gov.uk/independent_reviews/stern_review_economics_climate_change/sternreview_index.cfm (accessed February 2008).
8. ibid.
9. Greater London Authority 2007: "Action Today to Protect Tomorrow: The Major's Climate Change Action Plan," p.53, http://www.london.gov.uk/mayor/environment/climate-change/docs/ccap_fullreport.pdf (accessed July 2008).
10. London leading to a Green London: "Green homes concierge service," http://www.greenhomesconcierge.co.uk/ (accessed July 2008).
11.http://www.clintonfoundation.org/what-we-do/clinton-climate-initiative/our-approach/major-programs/making-buildings-green
12. Graham, S. and Marvin, S., 2001: *Splintering Urbanism: Networked Infrastructures, Technological Mobilities and the Urban Condition*, Routledge, London.
13. Monstadt, J., 2007: "Urban Governance and the Transition of Energy Systems: Institutional Change and Shifting Energy and Climate Policies in Berlin," *International Journal of Urban and Regional Research*, 31 (2): 326-343.
14. ibid.
15. The Merton Rule: The 'Merton Rule' is the groundbreaking planning policy, pioneered by the London Borough of Merton, which requires the use of renewable energy onsite to reduce annual carbon dioxide (CO_2) emissions in the built environment, http://www.themertonrule.org (accessed July 2008).

Chapter 5

Taking Action
Cynthia Rosenzweig

Cities, as home to over half the world's people, are at the forefront of responding to climate change. Climate extremes exert stress on urban environments through sea-level rise and storm surges affecting infrastructure, heat waves threatening the health of the elderly, the ill, and the very young, and droughts and floods threatening water supplies.[1] Though cities are vulnerable to the effects of climate change, they are also uniquely positioned to take a leadership role in responding to its challenges.

For cities around the world, the current task is to formulate Climate Change Action Plans that present a clear strategy for responding in terms of both mitigation of atmospheric greenhouse gas concentrations to reduce longer-term risks, and adaptation to current and future climate stresses in the shorter-term.[2] By addressing mitigation and adaptation on parallel tracks, city decision makers signal to their citizens and their wider network that both responses will be taken up simultaneously, avoiding the bifurcation that has often occurred, and assuring constituents that both short-term and long-term risks of climate change are being addressed.

New York City is addressing the climate change challenge through PlaNYC 2030, a program of its newly formed Office of Long-Term Planning and Sustainability. The plan includes multiple "entry points" for addressing reductions of greenhouse gases and for adapting to climate stresses.

Richard Plunz
There is a huge need for coordination of scientific efforts. The fragmentation of sciences is not doing anyone any good including science itself. The whole new environmental landscape that is emerging really lacks sufficient institutional recognition. For example we started a rather large study on childhood obesity. We have seen that the rise in childhood diabetes one could not have been imagined 20 or 30 years ago, and that has to do with many factors, including very large food production issues. We came to understand how much the obesity issue is connected with food production, and of course, with carbon footprint. And this comes without very much recognition on the science side because of the way research is structured. Climate change is no different. On

There is a huge need for coordination of scientific efforts. The fragmentation of sciences is not doing anyone any good including science itself

Climate Change Action Plans

Fundamental to formulating and carrying out a Climate Change Action Plan is that it is both a "process" and a "product." The essential process is stakeholder engagement, by which groups and individuals with a "stake" in the decisions are consulted at the outset and during the course of the activities.[3] These stakeholders include urban citizens and communities, city decision makers, and the private sector. Interactions with stakeholders are important for deciding on the balance of resources to be allocated to mitigation and adaptation, and for arriving at the specific response pathways for an individual city to implement.

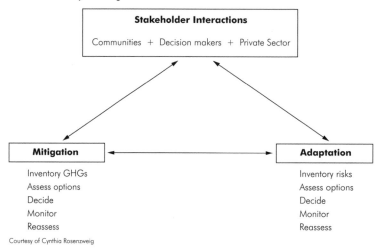

Courtesy of Cynthia Rosenzweig

Fig 5.1. Climate Change Action Plans for Cities

The steps for developing climate change mitigation actions to reduce the long-term risks include inventory, assessment, decision making, monitoring and reassessment. The first step is an inventory of the major sources of greenhouse gas (GHC) emissions. This step has already been undertaken by many municipalities through

---

the science side, there is an enormous context that we have to understand, and we don't have the institutional apparatus to do so.

Harriett Bulkeley

At the urban scale, I do want to question how much valuable science on climate change we actually have. A commonplace approach envisions different cities assessing gas emissions through an inventory, or assessing climate change risks, and then moving to another series of steps from assessment toward action. This model, in terms of emissions, has been around since 1993 with the worldwide "Cities for Climate Protection Program."* I would question how much action has come from this approach. I would also question whether now that we know,

At the urban scale, I do want to question how much valuable science on climate change we actually have

the Local Governments for Sustainability (ICLEI) process.[4] The next step is an assessment of the options for reducing emissions or for enhancing carbon storage, and a multi-criteria analysis of their costs, benefits, and implications. Urban designers play a key role here in creating solutions for these complex challenges. These steps then inform the decisions taken of which mitigation activities to implement in a particular metropolitan area.

Monitoring and reassessment are key aspects of the climate change action plans for several reasons. Since the global climate system is evolving and the exact trajectory that change will take is unknown, especially on regional scales, tracking global and regional indicators of both climate changes and responses helps to maintain flexibility and resilience. If changes such as sea-level rise appear to be occurring more rapidly than expected, both mitigation and adaptation actions can be revisited and upgraded. On the other hand, if changes appear to be slow in materializing, investments in costly responses may be delayed.

In parallel, steps for adaptation include an inventory of key vulnerabilities and impacts; an assessment of possible adaptation strategies; and a multi-criteria analysis of their costs, benefits, and implications. These steps in turn inform the decisions taken of which adaptation activities to implement. And just as with mitigation, monitoring of current regional and global climate trends and reassessment of adaptation strategies are critical to developing resilience to a dynamic climate that is likely to bring increasing heat waves, more droughts, more floods, and more coastal storm damage to cities around the world.

Role of Climate Change in PlaNYC
New York City is the most populous city in the United States, with

---

Perhaps learning the lessons from other cities would be more useful

basically, what emissions profiles in cities look like that we need to duplicate studies, we do not need to repeat the relatively expensive and time-consuming exercise of an inventory for each city when we know what will be found and what we have to do. It might have been a surprise for New York to find that their emissions came from buildings - but London already knew, and Melbourne has found the same, and Leicester has found the same, and other cities have found the same. So perhaps learning the lessons from other cities would be more useful. We know we have to reduce emissions from buildings and from transportation focused on urban contexts.

* The Cities for Climate Protection programme's "milestone framework" can be found at: http://www.iclei.org/index.php?id=810

a population of 8.2 million and a $60.2 billion operating budget.[5] A million more residents are projected by 2030.[6] In September 2006, Mayor Michael Bloomberg created the Office of Long-Term Planning and Sustainability, with the goal of developing a sustainability plan for the City. Responding to climate change plays a prominent role in the plan.

A Sustainability Advisory Board was formed of leading citizens with relevant backgrounds to guide the effort by identifying the major issues affecting the City's future, and working groups were created comprised of a broad group of experts in key areas including urban design, green buildings, climate change, and transportation. Mayor Bloomberg presented the goals of the sustainability program, PlaNYC 2030, in December 2006, and the actual plan in April 2007.[7]

PlaNYC includes specific goals in five areas – Land, Water, Air, Energy, and Transportation – with an overarching climate change goal to reduce global warming emissions by more than 30% below 2005 levels by 2030. Over 100 proposed actions are associated with the goals, including ways to reduce New York City's contribution to greenhouse gas emissions on the changing climate and also how to adapt to the projected climate changes in the city in the next two decades and beyond.

Embedding Climate Change in a Complex Urban Environment

The decision environment in regard to climate change responses in the New York metropolitan region is challenging, since it consists of intertwined jurisdictions of city, state, bi-state, and federal agencies. For example, New York State has several public authorities that are corporate instruments created by the legislature to further

---

Cynthia Rosenzweig
The world has probably spent billions of dollars to understand the climate system. But by comparison, this expense will be a very small fraction of the resources needed to understand the relationships between temperature change and associated climate impact. I will give you one example for food, which I have done a lot of work on. For agricultural production, we don't know the positive limits of the increasing carbon dioxide. We know that it is generally good for agricultural production because it increases photosynthesis, but at a certain point the negative effects of the higher temperatures globally shift the agricultural consideration from positive to negative, and we basically have no idea where that point is. For the world's food future, this understanding is something that

The world has probably spent billions of dollars to understand the climate system

public interests.[8] The Metropolitan Transit Authority (MTA) and Port Authority of New York and New Jersey are examples of public authorities that have significant roles to play in mitigation and adaptation to climate change in New York City. Complications in

Focus Areas	Goals
Land	
Housing	Create homes for almost a million more New Yorkers, while making housing more affordable and sustainable.
Open Space	Ensure all New Yorkers live within a 10-minute walk of a park.
Brownfields	Clean up all contaminated land in New York City.
Water	
Water Quality	Open 90% of waterways for recreation by reducing water pollution and preserving our natural areas.
Water Network	Develop critical backup systems for our aging water network to ensure long-term sustainability.
Air	
Air Quality	Achieve the cleanest air of any big city in America.
Energy	
Energy	Provide cleaner, more reliable power for every New Yorker by upgrading our energy infrastructure.
Transportation	
Congestion	Improve travel times by adding transit capacity for millions more residents.
State of Good Repair	Reach a full "state of good repair" on New York City's roads, subways, and rails for the first time in history.
Climate Change	
Climate Change	Reduce global warming emissions by more than 30%.

Table 5.1. Six Focus Areas and Ten Goals of PlaNYC 2030

is equally as important as the climate sensitivity to an acceptable carbon dioxide level.

Harriett Bulkeley
It is not so much the discovery of how precise a phenomenon climate change is, but rather the social function of raising awareness between politicians and the public about the importance of certain problems. It is no longer the scientist doing his own work in his ivory tower, but it is more and more advocacy science. So it is the scientist in the arena, the scientist asking sides. It seems to me that it may be – this is just my interpretation and I do want your judgment or your opinion - that we need not so much precision, but accuracy. What do I mean for example, that "precision" is? Will we have one degree or five de-

It is no longer the scientist doing his own work in his ivory tower, but it is more and more advocacy science

decision making and funding arise since New York State, New York City, and surrounding counties all share in the governance of these bodies.

Community involvement is also a key aspect of the decision environment in New York City, since there are numerous activist groups already in place that are focused on environmental justice and urban ecology. The community was involved in the creation of PlaNYC through website interactions and through town hall, neighborhood, and advocacy organization meetings. However, these early interactions were limited in time and opportunity.

PlaNYC includes a detailed structure for implementation, including identifying lead agencies and budget allocations. The oversight of each initiative is assigned to a lead agency. These lead agencies range from city agencies, such as the Department of Parks and Recreation, to joint agencies, such as the Metropolitan Transit Authority. Responsibilities that must be taken outside city jurisdiction are also identified. Additionally, future milestones are set by the years 2009 and 2015, such as "planting 15,000 street trees per year," and making projected budget allocations. The required investments by New York City in both capital and operating budgets are identified, together with other funding sources. However, funding the PlaNYC initiatives has proved challenging, since a significant portion of the New York City budget comes from New York State and state support is necessary for major new initiatives such as traffic mitigation through congestion pricing.

Progress and Prognosis
In April 2008, the Mayor's Office of Long-Term Planning and Sustainability issued a Report Card indicating the status of the many PlaNYC initiatives. In regard to climate change mitigation, they re-

---

grees of increase in temperature? Now many would say that a range between one and five is very likely. But of course this range is very imprecise, as the difference between the raise of 1°C and 5°C temperature changes the climate scenario completely. If we want to arrive at a point where we can say it will be precisely one rather than two degrees, then we have to invest a lot of money. My understanding from what you both have said is that many of the suggestions are commonsense suggestions. For sure they are based on science, on modeling, but at the end of the day, they are very accessible and reasonable things. For example, consider the impact of a heat wave: it is not only that you die because of heat, but also because there are invasions of species, new diseases, and so on. This idea of complexity, of interaction, requires

Many of the suggestions are commonsense suggestions. For sure they are based on science, on modeling, but at the end of the day, they are very accessible and reasonable things

port significant progress in transit-oriented rezoning, fuel-efficient yellow taxis, tree planting, reflective roofing requirements in new building codes, and rules allowing fuel-efficient micro-turbine generators that directly reduce the City's carbon footprint. Transport of solid waste out of the city has been switched from truck to barge and rail in Staten Island and the Bronx, and similar arrangements are being negotiated for the other boroughs.

On the legislative side, the City Council passed Local Law 55 in November 2007, which codified PlaNYC's goal of reducing citywide greenhouse gas emissions by 30% below 2005 levels by 2030. The law further requires the City to reduce carbon emissions by municipal operations at an even faster rate of 30% by 2017. Plans for implementing further reductions include avoiding sprawl and encouraging clean power, efficient buildings, and sustainable transportation.

In regard to adaptation, three initiatives have been launched. These are an intergovernmental task force to protect vital city infrastructure; the development of site-specific protection strategies with and for vulnerable neighborhoods; and a citywide strategic planning process for climate change adaptation. The intergovernmental task force will work with a technical committee of regional climate experts to develop coordinated climate protection levels for the metropolitan region. At the neighborhood level, two communities have been engaged – Sunset Park, Brooklyn, and Broad Channel, Queens. Feedback from these communities will inform a larger program of engagement with 40 particularly vulnerable neighborhoods throughout the city. The goal of the citywide strategic planning process is to update the Federal Emergency Management Administration (FEMA) 100-year flood plain maps.

---

firm understanding, requires the awareness that we are in a very complex system and so, a phenomenon like obesity is not only linked to food but it is as well linked to traffic, to urban planning, urban design.

Scientists have a definite advantage over philosophers, and maybe even over social scientists: they don't have to die to make their point

Antonio Navarra
We might say that scientists have a definite advantage over philosophers, and maybe even over social scientists: they don't have to die to make their point. Sir Thomas More had to die in the cathedral to show that his beliefs were important and supreme, but when Galileo Galilei was confronted with the alternative of either reneging his statement or going to the stake, he was able to save his life, because the earth is moving and nothing he could say would change this fact. We scientists have the ad-

PlaNYC encompasses a broad-ranging and challenging climate change agenda. Due to the short timeframe (~8 months) for development, community groups were not engaged fully in discussing the climate change adaptation aspects of the plan.[9] That is being remedied, at least in part, by the vulnerable community adaptation program now underway. On the mitigation side, a coincident Sustainability Task Force within the Metropolitan Transit Authority is providing a continuing forum for climate change discussions in the region, as is a Sustainable Buildings series presented by the New York Academy of Sciences. These activities highlight the important role that multiple local and regional organizations can play in engendering needed stakeholder involvement in responding to climate change.

UCCRN and IPC3

The Urban Climate Change Research Network (UCCRN) is a coalition of researchers from cities in both developing and developed countries, formed in 2007 at the time of the C40 Large Cities Climate Summit held in New York City.[10] A main activity of the UCCRN is the creation of the International Panel on Climate Change in Cities (IPC3) Assessment Report. The IPC3 represents a process by which research and expert knowledge may contribute to the development and implementation of effective urban climate change policies and programs. Since responding to the complex challenges of mitigation and adaptation requires a knowledge-based approach, the IPC3 report will provide a tool for policy makers as they "mainstream" response to climate change in urban areas. The IPC3 report will define both the "state of the knowledge" and the "state of action" in cities and identify key areas for further research relevant to the needs of urban policy makers. The initiating workshop for the first IPC3 Report will be held in the fall of 2008.

---

vantage that we are dealing with reality and I think we have to exploit the fact that there is an objective world and sometime we overstate the importance of communication. In the climate change issue, for instance, the true fact is that it is very difficult to change a way of life, especially when there are no incentives to do so.

The report will be divided into two parts: an overview of science and research and a city-by-city update of mitigation and adaptation programs and progress. The science and research overview will highlight key topics in climate science (including the urban heat island), mitigation, and adaptation addressed from an urban perspective. The report will cover the interactions between science and decision making, i.e., the significance and role of science to inform policies. Guidance will be given on city-specific methods and models for effectively incorporating science results into policy outcomes. Examples of effective programs will be highlighted as case studies. It will include discussion of social risks, economic risks, and physical, infrastructure, and ecosystem risks.

Conclusions and Recommendations

Because of their importance as commercial and cultural centers, cities have a leading role in climate change decision making. This role should be developed throughout both the public and private sectors, with strategic contributions played by knowledge providers and urban designers. Mitigation of and adaptation to climate change should be addressed simultaneously, with co-benefits explicitly highlighted. For example, bicycle paths and related exercise programs can result in reduction of greenhouse gas emissions from cars and buses, while helping to reduce obesity in often sedentary urban dwellers. Decision making should include participatory processes with stakeholders to ensure effective, relevant, and environmentally just actions. Here engagement of city communities and neighborhoods is essential. Finally, cities should engage with others in such initiatives as the C40 Large Cities Climate Summit, the Urban Climate Change Research Network (UCCRN), and the International Panel on Climate Change in Cities (IPC3) in order to develop coordinated and consistent preparedness. The IPC3 can play a key role in disseminating needed knowledge, tools, and guidance

Fig 5.2. C40 and UCCRN Symposium Representation

materials for use in developing climate change greenhouse gas and impact inventories, vulnerability assessments, multi-criteria analyses for proposed solutions, thus helping cities to carry out effective Climate Change Action Plans.

1. T.J. Wilbanks, P. Romero Lankao, M. Bao et al., Industry, settlement, and society. IPCC, 2007: Climate Change 2007: *Impacts, Adaptation and Vulnerability. Contribution of Working Group II to the Fourth Assessment Report of the Intergovernmental Panel on Climate Change*, Cambridge University Press, Cambridge, UK. 357-390.
2. M.T.J. Kopk and H.C. de Coninck, 2007. "Widening the scope of policies to address climate change: Directions for mainstreaming," *Environmental Science & Policy*: 587-599.
3. M.K. Van Aalst, T. Cannon, and I. Burton, "Community level adaptation to climate change: The potential role of participatory community risk assessment," *Global Environmental Change* 18 (2007) 165-179; and C. Vogel, S.C. Moser, R.E. Kasperson, and G.D. Dabelko, "Linking Vulnerability, Adaptation, and Resilience Science to Practice: Pathways, Players, and Partnerships," *Global Environmental Change* 17 (2007) 349-364.
4. http://www.iclei.org/
5. New York City Independent Budget Office, "Understanding New York's Budget: A Guide," 2005. http://www.ibo.nyc.ny.us/
6. Department of City Planning: Population Division, *"New York City Population Projections by Age/Sex and Borough, 2000-2030."* (New York, 2006)
7. Mayor's Office of the City of New York, "PlaNYC 2030," 2007. http://www.nyc.gov/html/planyc2030/html/home/home.shtml
http://www.nyc.gov/html/planyc2030/downloads/pdf/progress_2008_climate_change.pdf
8. New York State Office of the State Comptroller. "New York's Public Authorities." 2007. http://www.osc.state.ny.us/pubauth/index.htm
9. The author thanks Megan O'Grady for sharing her paper on evaluating community involvement in PlaNYC.
10. Urban Climate Change Research Network. 2008. http://www.uccrn.org/Site/Home.html

CHAPTER 6

The Adaptation Imperative
Richenda Connell

Carbon Neutral Is Not Climate Proof
The impacts of climate change are already being felt, and decisions based on historic climate data are no longer robust. The risks of dangerously high temperatures are increasing, along with greater incidence of flooding, strain on water resources and quality, and less stable ground conditions. What's more, however successful we are at reducing greenhouse gas emissions, further climate change is already built into the climate system for decades to come. Adapting to inevitable climate change is therefore now recognized as an essential part of ensuring our economic, environmental, and social systems can thrive now and in the future. Addressing climate change requires coordinated effort by both public and private sector decision-makers.

To date, climate change policies in both the public and private sector have tended to emphasise mitigation (that is, measures to control emissions of greenhouse gases) over adaptation. Yet, without an understanding of both climate change responses, there is a danger that actions taken to address one could thwart progress with addressing the other.

Complementary Actions on Mitigation and Adaptation
Considering urban development by way of example, mitigation and adaptation objectives are relevant at a range of inter-related scales – from building, to neighbourhood, to conurbation scales. Higher density developments have been promoted as a way of improving energy efficiency of urban areas. But if density is too high, it

---

Change becomes mandatory whereby climate change and economic responses to it are the only practical responsibility

Maria Paola Sutto
Perhaps at this point, the possible scenario can only be in one direction. When it comes to action in the face of climate change, the only possible choice for a city is to take action. The non-choice option, to do nothing, would only mean losing economic ground with respect to other cities who decide to take action. So, change becomes mandatory whereby climate change and economic responses to it are the only practical responsibility. The response to climate change is not a matter of choice, or of not being sure of which choice. Direct response is the only possible answer because the cities who will not be able to act will lose competitiveness, and will have to take second place to the others that will be making progress, pioneering directions that some are not bold enough to pursue.

can exacerbate the Urban Heat Island (UHI) effect and increase the likelihood of urban flooding. Recent urban adaptation research demonstrates these risks and shows how adaptation measures involving natural processes can be incorporated into high density de-

---

The "Adaptation Strategies for Climate Change in the Urban Environment" (ASCCUE) project

The ASCCUE project indicated that, even at high densities, well-designed urban green space and trees can provide shade and evaporative cooling, and can slow down runoff of surface water. Modelling studies in ASCCUE showed that a 10% decrease in urban green space could result in increased maximum surface temperatures in Manchester, England, of up to 8.2°C by the 2080s under a high greenhouse gas emissions scenario. But a 10% greenspace increase was shown to keep temperatures at or below current levels up until the 2080s.

---

velopments, providing multiple benefits.[1]

At the scale of an individual building, the need to address adaptation and mitigation synergistically is also apparent. The latest Intergovernmental Panel on Climate Change (IPCC) report, comments on the potential conflicts:

> In formulating climate change strategies, mitigation efforts need to be balanced with those aimed at adaptation. There are interactions between vulnerability, adaptation and mitigation in buildings through climatic conditions and energy systems. As a result of a warming climate, heating energy consumption will decline, but energy demand for cooling will increase while at the same time passive cooling techniques will become less effective.[2]

The report recommends that well-designed policies which "actively promote integrated building solutions for both mitigating and adapting to climate change" are vital if we are to prepare for the

---

Harriett Bulkeley

One of the things that is striking about the London case, so far, is that despite the magnitude of the target of taking London off the grid, of reducing emissions by 60% by 2030, there has been relatively little opposition. There doesn't seem to be an economic group who feels frightened by this. This shows the power of leadership, in that the Mayor has put to the forefront the potential for the economic fix, and he is trying to get other leading proponents of businesses to sign up, in particular the financial sector.

Despite the magnitude of the target of taking London off the grid, of reducing emissions by 60% by 2030, there has been relatively little opposition

Richenda Connell

In London we have an organization called the London Climate Change Partnership, which is a mixture of public

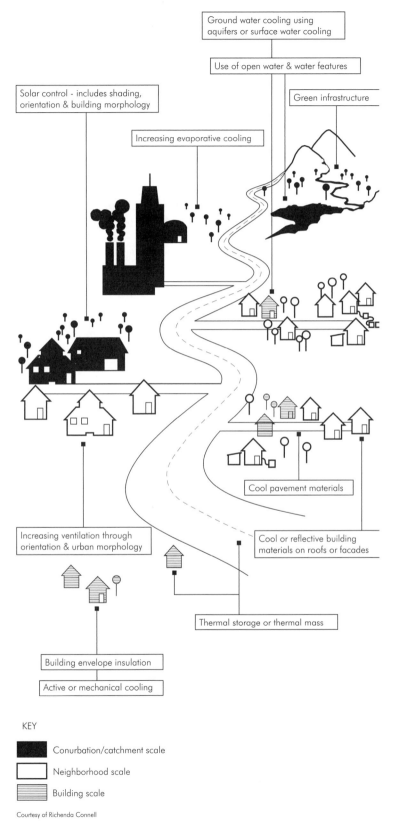

Ground water cooling using aquifers or surface water cooling

Use of open water & water features

Green infrastructure

Solar control - includes shading, orientation & building morphology

Increasing evaporative cooling

Cool pavement materials

Increasing ventilation through orientation & urban morphology

Cool or reflective building materials on roofs or facades

Thermal storage or thermal mass

Building envelope insulation

Active or mechanical cooling

KEY

Conurbation/catchment scale

Neighborhood scale

Building scale

Courtesy of Richenda Connell

Fig 6.1. A Menu of Strategies for Managing Flood Risks

impacts of a changing climate without compromising emissions reduction targets.

Progress is being made in bringing this integrated climate risk management approach into decision making by built-environment professionals. New guidance aimed at planners and designers is now emerging, demonstrating how spatial planning and urban design at all scales can contribute to ensuring that urban areas are well adapted.[3] For example, Figure 6.1 provides a menu of strategies for managing flood risks.

Governments Are Beginning to Promote Adaptation

Strategies for adaptation are increasingly addressed in formulating public policy. At an international level, the United Nations Framework Convention on Climate Change (UNFCCC) refers to adaptation in Articles 2 and 4. Under the Framework, the Nairobi Work Programme on impacts, vulnerability, and adaptation to climate change was launched in 2005.[4] The objective of the five-year programme is to help all countries improve their understanding of the impacts of climate change and make informed decisions on practical adaptation actions and measures. In 2007, the Bali Action Plan identified the need for enhanced action on adaptation by parties to the Convention.[5]

In Europe, adaptation has been progressed by the European Climate Change Programme II and the recent EC Green Paper on Adaptation.[6] The Green Paper argues both for mitigation and adaptation and describes possible avenues for action at EU level. Its main objective is to kick-start a Europe-wide public debate and consultation on how to take it forward.

National adaptation plans have also been produced and many coun-

---

and private sector organizations and involves some of the best climate change researchers working in London. The partnership's financial services group has been looking at the risks that climate change poses to the financial success of the City, and the financial services companies in the City are looking to the Mayor of London, saying "we want to know that you are providing us with a climate-resilient environment in which to be based." At the same time the Mayor of London is saying, "I want you to stay here. I don't want you to move to another city because London isn't offering the environment you need." So everyone recognizes they need each other and they need to work together to address adaptation in a sustainable way.

The partnership's financial services group has been looking at the risks that climate change poses to the financial success of the City

tries are now moving to develop and implement adaptation policy frameworks.[7] Adaptation plans are also being developed at regional and local government level.[8]

Where Does This Leave Business?

The issue of participation of the business community leaves much to be desired. Our experience of working with businesses on prioritizing adaptation over the last three years reveals the following explanations for this situation:
• the greater emphasis placed by governments on the mitigation agenda
• the perception that climate change is too far in the future to be relevant to today's business decisions
• the perception that information on climate change is too uncertain to be included in business decisions
• poor understanding of how climate change will affect most business sectors – most research effort has focused on impacts on natural systems and does not demonstrate the "business case" for adaptation
• many public sector regulations, contracts with the private sector etc, do not yet require adaptation.
• some businesses do not recognize climate change adaptation as a mainstream business risk issue; rather they perceive it as a "green" or corporate social responsibility (CSR) issue.

Yet failure to understand climate change risks to assets and portfolios could potentially create liabilities for companies. As the financial impacts of climate change begin to be more widely recognized, litigation could be used as a means to recover costs. Indeed, lawyers are beginning to acknowledge that there is sufficient information available on climate change for companies to take it into account in both strategic and project level decision making. For instance,

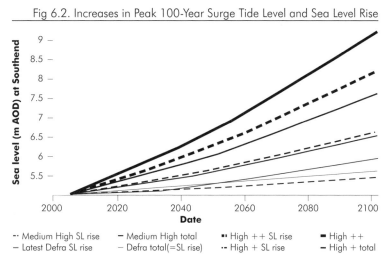

Fig 6.2. Increases in Peak 100-Year Surge Tide Level and Sea Level Rise

Sea level (m AOD) at Southend

9 –
8.5 –
8 –
7.5 –
7 –
6.5 –
6 –
5.5 –

2000   2020   2040   2060   2080   2100

Date

-· Medium High SL rise   — Medium High total   ▪· High ++ SL rise   ▪ High ++
— Latest Defra SL rise   — Defra total(=SL rise)   ·· High + SL rise   — High + total

Source: Acclimatise and Climate Risk Management Limited 2008

in relation to adaptation of the built environment, a leading UK property law firm has stated:

> The effects of climate change can now be regarded as being reasonably foreseeable and at every stage – from initial instruction, through the design and planning process to construction and beyond – it must be incumbent upon professional advisers to ensure that appropriate steps have been taken.[9]

The potential for litigation around climate change is a serious issue. Decisions taken by directors and professional advisers that do not take climate change into account may be open to legal challenge. In the future, the courts will examine claims and may decide that it was reasonable, when the decision was made or advice given, to have foreseen the impacts of climate change, based on the information available in the public domain.

Addressing climate change adaptation is beginning to be recognized as a matter of good corporate governance and fiduciary responsibility. Company directors are being challenged by investors to demonstrate their corporate governance credentials and improve their disclosure of future risks. Within this context, a recent report by a group of pension funds and fund managers calls for engagement and dialogue with companies:

> to ensure that they recognise climate change adaptation as a risk, that they explain to investors how they have assessed these risks, and that they have established appropriate systems and processes to respond to these risks.[10]

What Steps Could Businesses Take?
As a first step, businesses can undertake risk assessments of their business models, to understand the impacts and costs of a changing climate. For instance, climate risks can be incorporated into cash

---

In the UK, another reason for climate change getting very real is that standard household insurance policies cover flood risk, and people who have been flooded have seen their insurance premiums increase to several thousand pounds a year. It is getting really, really expensive. Properties cannot be sold, hitting homeowners in a very personal way and making the issue of climate change very real. The issue for me is how we get people to be proactive, not reactive, in terms of adapting to climate change. Politicians only seem to react to extreme events. How do we get them to act proactively, before the disaster strikes?

Climate risks can affect all aspects of a business model

• Raw materials: Increasing price volatility in raw materials can be expected.

• Third party utility, transport, and communications suppliers: Companies will be affected by climatic impacts on their suppliers. Securing long-term relationships with suppliers who are themselves "adapted" will become increasingly important.

• Design and location of new assets: Engineering designs based on historic climate information will likely create future failures, costs, and liabilities.

• Operational impacts on existing processes: The operating margins, thresholds and sensitivities of existing assets, plant, and equipment could be compromised.

• Markets and customers: Customer preferences and needs may change in response to climate change. Major changes to some markets can be expected over time.

• Opportunities to develop new products and services: There will be winners and losers from climate change; the winners will likely be those companies that are early to recognize the importance of climate change, foresee the implications for their sector and their customers, and take appropriate steps before their competitors.

• Workforce: Changing health and safety legislation in response to climate risks could become a major issue.

• Interplay between a company's operations, local communities and the environment: Conflicts could arise - for example, greater competition for water in areas that become more water-stressed.

• Reputational risks: A company's brand could be harmed through failure to address climate change - for instance, if flood risks or high temperatures made property developments unattractive or unusable.

• Stakeholder perceptions and actions: Businesses will be affected by the adaptation strategies, policies, and regulations being developed by governments, as well as the adaptation strategies of their investors, bankers and insurers.

flow forecasting, examining impacts over time on operating and capital costs and on income. Some of the key risk areas to be considered are outlined in the box above. Businesses that are highly dependent for their success on large fixed assets, or on raw materi-

Fig 6.3. Weather Related Catastrophes and Insured Losses

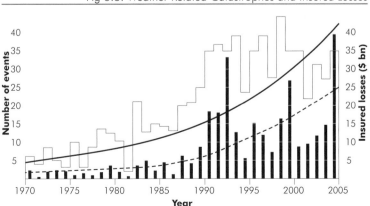

Weather-related events   — Trend (Weather-related events)
■ Annual insured losses   -- Trend (Annual insured losses)

Source: "Financial risks of climate change" ABI, 2005. Report prepared by Acclimatise

als and complex supply chains, are particularly vulnerable. Having assessed the risks and costs they face, businesses can then begin to develop suitable adaptation strategies and to mainstream these across the structures of the organization.

1. S. Gill, J. Handley, R. Ennos, and S. Pauleit, "Adapting Cities for Climate Change: the Role of the Green Infrastructure," *Built Environment* 33, no. 1 (2007): 115–133.

2. IPCC, 2007: *Climate Change 2007: Mitigation. Contribution of Working Group III to the Fourth Assessment Report of the Intergovernmental Panel on Climate Change,* Cambridge University Press, Cambridge, United Kingdom and New York, NY, USA.

3. R. Shaw, M. Colley, and R. Connell, "Climate Change Adaptation by Design: A Guide for Sustainable Communities," London: TCPA, 2007. http://www.tcpa.org.uk/downloads/20070523_CCA_lowres.pdf (accessed July 2008).

4. UNFCCC 2008: *Nairobi Work Programme on Impacts, Vulnerability and Adaptation to Climate Change,* http://unfccc.int/adaptation/sbsta_agenda_item_adaptation/items/3633.php (accessed July 2008).

5. UNFCCC 2008: *Decision-/CP.13, Bali Action Plan, Advanced Unedited Version,* http://unfccc.int/files/meetings/cop_13/application/pdf/cp_bali_action.pdf (accessed July 2008).

6. EU Commission 2007: *Adapting to Climate Change in Europe – Options for EU Action,* Green Paper from the Commission to the Council, the European Parliament, the European Economic and Social Committee and the Committee of the Regions.

7. UNFCCC 2005: *Vulnerability and Adaptation to Climate Change in Europe,* EEA, Technical Report No 7/2005, ISSN 1725-2237, http://reports.eea.europa.eu/technical_report_2005_1207_144937/en/EEA_Technical_report_7_2005.pdf (accessed July 2008).

8. UK Climate Impacts Programme 2008, http://www.ukcip.org.uk/ (accessed July 2008).

9. M. Dowden and A.C. Marks, "Come Rain or Shine," *Estates Gazette,* July 16, 2005.

10. Henderson Global Investors, 2008, *Managing the Unavoidable: Understanding the Investment Implications of Adapting to Climate Change.* Universities Superannuation Scheme, RAILPEN Investments, Insight Investment. http://www.insightinvestment.com/Documents/responsibility/Reports/Managing_the_Unavoidable_Understanding_the_investment_implications_of_adapting _to_climate_change.pdf (accessed July 2008).

CHAPTER 7:   **Urban Competitiveness**
Matteo Caroli

The Challenge of Climate Change

Due to global competition, the mobility of capital flows, and technological innovations, policy makers are starting to refocus their attention toward cities to ensure their international competitiveness. This can be accomplished through improving economic and social conditions. Within this perspective, cities are competing directly against each other, and so they map out their territorial strategies to promote their own development. At the same time, it is crucial to understand that from a territorial standpoint, the desire to compete should come with conditions. The notion of "sustainable competition" among cities should guide their policy and planning initiatives within this renewed international perspective. Global warming, however, threatens to disrupt the ideal of "sustainable competition."

The ability of a city to compete is not based on acquiring a better standing within the global competitive hierarchy. Rather, implied is a plethora of "softer" goals which, if reached, allow the city to pursue a certain path of development that reinforces its own success from a planning, policy, and economic standpoint. Development goals stem from various contingencies. Among the most pressing is the issue of climate change, which is a factor that must be deeply analyzed and managed by the relevant economic players. The mitigation and adaptation of climate change has in recent years become the new paradigm for managing and planning territories and cities.

Policy makers, whose main goals include achieving a certain degree

---

Bruna De Marchi
The language of mainstream economics is not the only language that we need to speak. It is one of the languages. It is one area of knowledge. It is one of the sciences, but there are others.

I think the required knowledge will not come from economists. It will not come from politicians, because they have to win the vote

Lieven De Cauter
I think the required knowledge will not come from economists. It will not come from politicians, because they have to win the vote. I don't know from whom it will come. I think from civil society asking, pleading, pressuring that we take action. Immediate action.

Claudia Bettiol
Any kind of project financing initiative is based on the fact

of competitiveness, should be aware that climate concerns represent an obligation as well as an opportunity for further growth within cities. On one hand, they limit the ease to manage territorial resources; but at the same time, they present the opportunity of promoting very advanced and sustainable development. In learning to manage the climate change issue, cities can strengthen their abilities to manage external risks, and even to transform them into opportunities for economic and social gain. In the light of these considerations, it is useful to develop a paradigm of "sustainable" or "harmonic" competitiveness, which would help to implement the most efficient development strategies.

Competitiveness as Resource Availability

Territorial competitiveness expresses itself in a geographical area's material and immaterial resources, which are functional to its harmonic development. The concept of "harmonic development" is used to highlight that the three traditional facets of sustainability, the economic, social, and environmental dimensions, must be managed within a general equilibrium, without favoring any particular dimension, as it would be to the detriment to the others and to the system as a whole. Resources that are functional to an area's harmonic development are those that impact the competitiveness of the economic organizations and activities in addition to human capital located within the territory; and that positively contribute to its harmonic development. Therefore, the reinforcement of a territory's competitiveness is dependent on the enrichment of this set of resources. A territory competes with the aim of gaining and developing the best set of resources for its harmonic growth.

Resource Accumulation Within a Territory

It is crucial to clearly define the term "accumulation" so as to get a better understanding of its implications with regard to the activities

---

that I don't have an immediate advantage in undertaking a specific activity, and therefore is diluted over time because I draw advantages from the use of that building or whatever facility. So, if we apply these advantages to the urban planning scale, I am linked to a negotiation that is diluted over time, and the identification of the time frame is the difficult part.

Harriett Bulkeley
The issue relates to competitiveness. In London if you ask why the City decided to act as it has, the Mayor's office will tell you that their public opinion polling is putting climate change in the top three or four issues for the City. For companies it's becoming an increasingly important issue, related to decisions about whether they locate in

Any kind of project financing initiative is based on the fact that I don't have an immediate advantage in undertaking a specific activity, and therefore is diluted over time

that generate and develop a stock of resources. The five activities that generate and develop the stock of resources over time are the following:

• Building - so as to favor the conditions of the resource generation within the territory.

• Attracting - so as to stimulate the enrichment of the original set of internal resources through external ones, and to strengthen the conditions for its constitution.

• Improving - so as to individuate the best possible conditions to exploit resources from a sustainable development framework.

• Maintaining - so as to encourage the embedding of potentially mobile resources, as well as their integration within the local economic and social system.

• Renewing - so as to activate the resources' renewal mechanisms, in line with the evolution of competitive scenarios.

Territorial Resources that Attract

Given the set of resources that are directly or indirectly available, the territory should be the most preferential place for organizations to locate to and be involved in prosperous economic activities as well as be the "best" place to settle and fulfill one's professional and personal life. The stock of territorial resources reflects a capacity to attract organizations, economic activities, and people. Territorial "competitiveness" expresses itself in its "attractivity."

The term "attractivity" in this context should be interpreted in an expansive way, so as to include the attraction within the territory of external agents, as well as the maintenance and growth of those agents already located within the territory itself. Depending on the specificities of the set of organizations, people, and economic activities embedded in the territory, as well as the socio-environmental conditions, the pattern of harmonic development is determined

Fig 7.1. Drivers of Urban Competitiveness

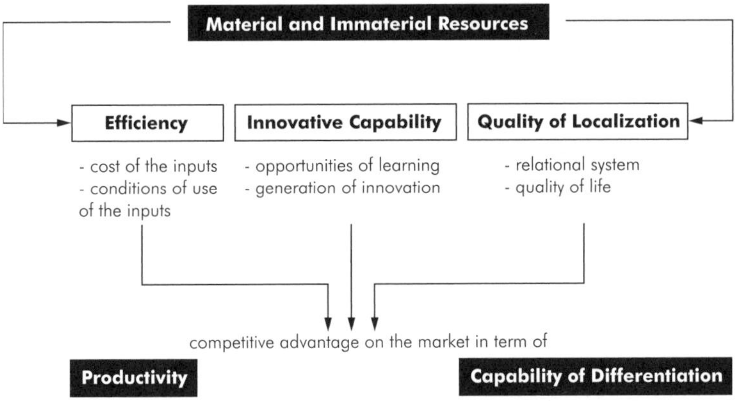

A city may build its competitive advantage on what we can call those "locally-specific and embedded nontradable assets and endowments" (Martin, 2005)

Courtesy of Matteo Caroli

and the territory evolves.

People, Organizations, and Economic Activities

In a sustainable development framework, people, organizations, and economic activities have more than one value. On the one hand, they originate a specific path of territorial harmonic development, and so they represent the subject of its attraction strategy. On the other hand, they are crucial components of the territorial stock of resources and are, therefore, drivers of its attractivity degree. Thus, people, organizations, and productive activities are resources, as well as generators of resources.

Territorial competitive strategy must be based on the types of economic organizations and people that inhabit the territory. Consequently they determine how the territory gets defined according to certain characteristics as being the "best place" to settle. It is crucial to remember that no territory may aspire to be the "best place" categorically for many different types of economic organizations. Political leaders may take the lead in deciding on a specific course of sustainable development strategy, though they still will be strongly influenced by the territory's original stock of resources as well as its path of dependency.

A territory is composed of divided parts that compete against each other and result in winners and losers. Territories also compete against one another with resulting ranking placements. A sports metaphor may be useful. Competition among territories is similar to a golf contest in which each player must improve their starting "handicap." There is an important exception to this principle: territories directly compete against each for the acquisition of scarce resources as well as for the economic organizations that greatly contribute to harmonic development, provided that their supply is

---

London or somewhere else. There is a kind of "green competitiveness" happening.

Claudia Bettiol
The policy makers are the real decision makers, or more precisely, the financial policy makers are making the real decisions. So far we have always talked about policy decision makers but actually we have forgotten the fact that policy makers are servants of the financial world. The financial world is finding some benefits in advocating for the environment, starting with the insurance and re-insurance sector. Three years ago I participated in a conference in Milan on energy problems, convinced that I would find many engineers in attendance. Only a small percentage of the participants were engineers, however.

The financial world is finding some benefits in advocating for the environment, starting with the insurance and re-insurance sector

greater than the demand.

Territorial demand is defined by the organizations and the economic activities by which a territory strives to be the most attractive with respect to alternate locations. Territorial competitiveness is expressed in its ability to satisfy demand better than other territories. Once again, this ability is determined by the quality and composition of the territory's stock of resources.

Territorial Competitive Advantage: Distinctive Resources
The composition and quality of territorial resources determine any "competitive advantages" with respect to alternate territories. This competitive advantage must be "sustainable." Not all of the territory's available resources are relevant to the creation of a sustainable competitive advantage. For example, there are certain resources that constitute primal elements that spur on the development of productive activities.

"Distinctive" resources are those that a territory holds exclusively or on a superior level. Among the territorial resources, it is important to select those that have a "distinctive" nature from competing territories, and to consequently promote mechanisms that further attractivity. The distinctness and usefulness of a resource can be seen when considering a specific demand. For example, a unique natural landscape is a distinctive resource when considering tourist demand, but not for investor demand from the steel industry. Many territories have "threshold" resources, and may not hold distinctive resources. There are however limitations to this perspective. On the one hand, the stock of resources constitutes only a potential driver of the territory's attractivity. The way in which it manifests itself depends on the way in which resources are actually utilized and improved upon. Moreover, it is crucial to consider the resources'

If we are all already convinced that sooner or later consequent disasters will strike, the problem becomes what are the first five things to do after the disaster

Most were financial people - economists. The problem is not so much alarmism, or the lack of alarmism, or what we're going to do in order to face the catastrophe. If we are all already convinced that climate change is under way, and if we are all already convinced that sooner or later consequent disasters will strike, the problem becomes what are the first five things to do after the disaster. We must ask not only how do I build the house, but what are the first five political actions. For example, if the Mayor of New York city has thousands of people's homes under water, what are the five political actions he must put in place?

Harriett Bulkeley
Climate leadership entails different issues. How you en-

"capabilities" in terms of its utility as well its flexibility to build upon and improve the existing structures within the territory.

On the other hand, the stock of available resources constitutes "static" information that does not represent the quality of the territory's evolution with respect to the evolution of the demand as well as of the more general conditions on which its harmonic development depends. To address this limit, it is important to consider the territory's "change capability," which refers to the capability to activate forces which can shape the stock of territorial resources to more coherently evolve with the changes of the context conditions as well as the characteristics of demand.

Accumulation Capability of Resources

In order to understand territorial competitiveness, it is crucial to evaluate not only the accumulation of resources within a timeframe, but also to be able to substantiate the process by which the accumulation happens. The way in which the resources are accumulated within a timeline attributes a factor of uniqueness as well as "stickiness" to the territory. In the current globally competitive environment, in which the mobility of resources is particularly efficient, the "sticky" component of the resources' stock has competitive value. To this end, the "intangible" components of the territorial resources are extremely relevant. Moreover, it's important to be able to renew the stock of available resources so as to balance the intrinsic resources' mobility along with the relative ease of imitation.

The importance of the "accumulation" process has a relevant implication: territorial competitiveness depends indeed not only on the stock of resources, but also on its ability to build, attract, enhance, maintain, and renew the most appropriate resources for its sustainable development. In order to be competitive, a territory

---

gage, through leadership, an idea of co-benefits. For example, you can promote the issue of climate change and obesity - of more liveable, walkable cities. You can promote the issue of climate change and energy - the benefits to householders who have installed photovoltaic panels and are actually generating energy for the grid rather than consuming it. These co-benefits work more at the household or the neighborhood scale, rather than at the city scale as a whole.

must hold the factors that generate this ability.

This ability can be attributed to some of the territorial resources or it may be the outcome of specific actions implemented by certain actors within the territory. For example, an internationally qualified university exemplifies both attributes of resource generation: its scientific and formative activities not only are an attraction factor for high-level students and researchers, but they also determine their continued renewal and favor their embedment in the economic and social system of the territory itself, at least for a certain period of time. A well-designed foreign investment's attraction planning, implemented for the long term, is a clear example in which a territory may develop its ability to constitute, attract, enhance, maintain, and renew its resources.

Determinants of a Territory's Attraction Capability

A territory's attraction capability is determined by two conditions: the quality and composition of its stock of resources, in particular, of its "distinctive" resources; and the capability to accumulate resources, in particular, "distinctive" resources. The indicators that survey the stock of distinctive resources must be able to detect their "distinctivity." It's important to remember that "distinctivity" is present when a territory holds a resource in an exclusive or relatively superior way. Consequently, each indicator must be designed with respect to the average value determined within a set of considered territories, or in comparison to the total value.

Choice of Target Demand

It is important to correctly assess the territory's "demand" with relation to its supply. A crucial step of a territory's competitive strategy is market positioning. The target demand should be chosen with respect to two main principles: the contribution each demand

Fig 7.2. Economic Competitiveness and Cities' Attractivity

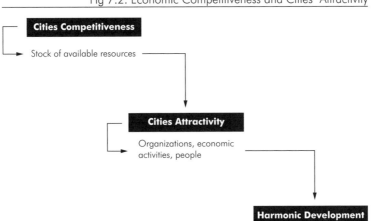

Courtesy of Matteo Caroli

category may provide to the sustainable development of the city; and the assessment of how well the territory's resources satisfy the needs of each demand category.

In general, a territory is successful when its policies and actions reflect an equilibrium between the demand targets and its supply of available resources. An ability to improve upon resources must also be present. When assessing demand categories, it is important to remember that sometimes they may only be partially defined, and consequently, this may contribute to some ambiguity, especially in so far as climate change is concerned.

CHAPTER 8

What If . . . New York
David Burney

New York City was very much affected by the impact of Hurricane Katrina in 2005 and is aware of the probability of some similar event happening in New York in the future. One of the lessons learned from Katrina was the importance of emergency housing, which was dealt with somewhat haphazardly in the New Orleans and Louisiana with the use of trailer-type mobile-homes supplied by FEMA – Federal Emergency Management Agency. The trailers attained only very low density housing. As a result, New Orleans lost more than half of its population because they could not provide enough trailers quickly enough, or in sufficient density on the available, non-flooded land to meets the needs of the population. The economic implications for the city have been very serious. As a result, it is possible that New Orleans will never fully recover - it will almost certainly never recover its population and may not recover economically.

In response to this problem, the OEM, the acronym of the Office of Emergency Management of New York City, which has the responsibility to plan for emergency eventualities, be they attacks or floods or earthquakes or whatever, launched a competition just recently. OEM has planned for force 3 and 4 hurricanes with a series of measures to evacuate the most seriously affected areas, provide emergency shelter, food and water, medical care, communications, and other immediate post-disaster measures. The purpose of the competition was to investigate what the City can do in response to flooding as a result of a hurricane, through provision of emergency housing for the somewhat longer period of recovery and

---

We need to design a post-catastrophe scenario addressing how to modify decisions and how to help decision makers to become separated from the financial world

Claudia Bettiol
We need to design a post-catastrophe scenario addressing how to modify decisions and how to help decision makers to become separated from the financial world. The multinational companies, coming from Asia, India, Russia, China and United States, have changed the decision-making process in such a radical way that a catastrophe can actually restore some power to politicians. And we really need to talk about what they must do in a post-catastrophic situation.

Massimo Alesii
I would completely change the approach. We have learned what post-catastrophism is. We have learned in the last 30 years that populations with "zero risk percep-

reconstruction of the affected neighborhoods. Emergency housing would be needed quickly and economically, and at a density that matches the level of density that exists in the City. Rather than ten households per acre in trailer homes, is it possible to design another form of provisional post-disaster housing, closer to the 200 households per acre in our estimates, needed to retain our population? Competitors were invited to submit proposals in response to that question.[1]

There were 117 entries to the competition. Interestingly enough, 62 of them – by far the greatest number, were from the United States. The next largest contributing country was Italy, thanks to its wealth of design talent -and after that, from many other places around the world. The city selected ten competition winners and paid a stipend to those ten winners to develop their ideas further so that, ultimately, we will have several potential scenarios that we can plan for and perhaps even prefabricate and be ready should the need arise.

Scenario of the Competition

There have been hurricanes in the past affecting New York City, some of them fairly devastating, and climatologists believe that in the future they will increase in frequency and force. New York City, historically, has been blessed by a well-protected harbor, and grew partly because of the benefit of that harbor. But now, with coastal storms, flooding is more likely because of the confluence of rising tides and storms forcing water up the coast, combined with the Hudson River flowing down toward the ocean past Manhattan. The potential is increasing for surges in water levels all around the city - not only on the coastal areas, but also inland further up the estuary. The more exposed areas of the Far Rockaways and the Jersey Shore would be affected first, but followed by a far larger area comprising 2.3 million people in a Category 3 or 4 coastal storm. As water

---

tion", as in the case of Chernobyl, suffer the maximum catastrophic impact.

Richard Plunz
To me, the real milestone of the "What if..." competition was the positive example whereby New York City gave public acknowledgment of the possibility of an extreme climate event, and began to talk responsibly about what to do, about how we would prepare. Public knowledge is crucial in building toward public engagement. I will be curious to see what happens with "What If?" For example, will the media accord this competition the importance that it deserves? Will the design professions give the design results importance, especially given that they are not design in the "high design" sense; not in the sense

Public knowledge is crucial in building toward public engagement. I will be curious to see what happens with "What If..."

levels rise, obviously most affected would be those who are nearer the ocean side, but also many others in Manhattan and even up as far as the Bronx.

The design competition was one portion of a broader strategy to provide provisional post-disaster housing. In its planning for a coastal storm, the OEM developed a strategy for moving people to temporary shelters. Staging would happen the day before the storm, involving use of evacuation routes and the shelter systems, based on the projected inundation levels. Flood levels can be predicted based on computer modeling. The Far Rockaways might expect 3.6 to 7 metres inundation or increase in water level. Even farther inland we can expect 1.8 to 3.6 metres, and the least affected areas would still see 1.0 to 1.8 metres increase. And this is only in the Category 3 hurricane, which is highly probable, but there is also the possibility of a Category 4 hurricane where inundations would be even greater. For the purposes of the competition we assumed Category 3.

The emergency strategy includes forecasting of the storm, broadcasting to residents, communication, and evacuation. OEM would be responsible for organizing the response using trained volunteers, alerting city services like fire, police, sanitation, and Red Cross services.

New York City has already identified temporary shelters in schools that are outside the flood zone, where people will be relocated and emergency services will be provided. After the storm, rescue operations would ensure that people are out of the inundated areas. Restoring power becomes priority number one because very little can be done until the power system is restored. The City's sanitation workers and the Army Corps of Engineers are brought in to

that the design culture is used to operating.

Lieven De Cauter
I am suggesting for New York that we volunteer to do a trial exercise, and taking Julie Sze on board, give New Yorkers not only these sort of house kits, but also survivor kits like "how to survive without power for 50 days." Because, indeed, if you have your house kit, that is one thing; but you also need quite a scenario for daily routine, an algorithm that people can live by.

If you have your house kit, that is one thing; but you also need quite a scenario for daily routine, an algorithm that people can live by

David Burney
We can certainly challenge some of the ideas that underlie the competition. We started with the idea, for example, that a flood is a one-off event that happens and goes

handle the debris, from roads in particular, because those needed to get access to inundated areas. Then assessment begins: how many buildings are completely destroyed and need to be demolished, how many need to be repaired and so on.

The shelter systems are then transferred to Red Cross supervision who will take over the care of people who cannot return to their homes. And then transportation services start to come back, beginning with buses and then subways. Sites begin to be cleared for temporary housing as debris is relocated on barges for disposal. By day 80 after the storm, cell-phone coverage is restored (I don't know how New York can manage to last 80 days without cell phone coverage but this is the plan! We'll see if it works!). And then, of course, it starts to get very cold; because hurricane season is followed by the very cold New York winters. And we are left with the possibility of 20 thousand or more households displaced. This is the sort of target population for the temporary housing. And by now there are potential sites where houses can be sited, roads giving access to these sites, and there is power restored to enable them to be put in place and hooked up to utilities.

The Competition Judging Criteria
• High Density - Most important and the thing that makes this competition for disaster housing perhaps different from others that have taken place in other parts of the world.
• Rapid Deployment - Units ready to be occupied as soon as possible.
• Site Flexibility - Maximize the ability to accommodate as many different types of sites as possible.
• Unit Flexibility - Maximize the ability to accommodate as many variable household types as possible.
• Reusability - Maximize the potential for reuse of the structures

---

away. But maybe it is a multiple event. If so, what impact would this have had on the outcome of the competition? For example, is the assumption that flooded areas will mostly recede and could be re-occupied really a sound assumption? More importantly, the competition should be a vehicle for the alarmists, a wake-up call to the citizens in New York, to the administrators of New York, to the general population, and to the denial community. We very purposely did not make the site of the competition an abstract site; it is a real physical place in the city, and we were hoping that the media would say to people: look, this is actually, really, what could happen. And there is a good probability that it will happen, so don't just ignore it. Embrace it and think about its implications. To me, this would be the most successful outcome: an awaken-

The competition should be a vehicle for the alarmists, a wakeup call to the citizens in New York, to the administrators of New York, to the general population, and to the denial community

for either future disasters or other purposes.
• Livability - Maximize the strength, utility, convenience, and comfort of the dwellings.
• Accessibility - Allow access for people with limited mobility.
• Security - Make public space defensible and help people feel safe.
• Sustainability - Reduce energy costs and the carbon footprint of the dwellings.
• Identity - Maximize the ability of New Yorkers to feel a sense of identity and pride in where they live.
• Cost Efficiency - Maximize the value of the investment.

In the competition submissions that were received, a very popular choice was shipping containers. They are standard, widely available units that are appropriately sized for conversion to habitation. In many places there are examples of these containers being adapted for habitation in a someway ad hoc way, by individuals informally cutting holes for windows and so on. They are widely available, inexpensive, and can be transported by ship. Most importantly, they can be stacked up to six or seven high in order to achieve the necessary density. Many competitors used this model as a way of delivering temporary housing very quickly. One particularly interesting strategy was to put the temporary houses on the streets rather than on cleared vacant lots. It assumed that a typical New York City street, with three- or four- story houses, might be too severely damaged to be inhabited. So temporary trailers were proposed to be put on the street, and that enables work to be undertaken - either demolishing or repairing the damaged buildings while the population remains in the neighbourhood.

A number of entries used the typology of the container module, but custom-designed from scratch. One interesting example came

---

ing that might help breach that gap between theory and practice, between leadership and consensus, and really get us focused, as a city, on the challenge that's obviously out there.

Cinzia Abbate
It is quite interesting – as an architect - to see the gap between the questions about designing temporary housing, and the fact that the scientists are saying, "What if this event happens many, many times?" As a designer, I would say we have different issues: one is the housing, another one is the issue of the infrastructure. In fact following this logic, the post-catastrophic land is not any longer a land to be inhabited. The topography has changed completely and the tools that we have to use for designing or plan-

Following this logic, the post-catastrophic land is not any longer a land to be inhabited

collapsed into five little green packages that included panels for the roof and walls, the bathroom and utility storage, kitchens, wall and glazing panels and then a floor. They can be unpacked and deployed into shelter modules. They can be stacked up to four stories in height, and are flexible enough to occupy spaces between buildings so that damaged buildings can either be repaired or replaced. Another idea involved a delivery truck that brings little habitations in containers. The trucks transform themselves into gantries, which are placed along the streets and used to support the dwelling units above the street. Again, people can inhabit the street while the buildings are being demolished or repaired.

Another approach was to deploy a hybrid container. In order to make it easier to store and transport, it would be partly rigid and partly expandable. So for example, there could be a cube and the cube would be transported, but 50% of the unit is in a flexible fabric or other material that would expand once the cube is stacked into place. There were also a number of entries that used hexagonal geometry to create a kind of "honeycomb" approach to the problem, similar to the beehive. In one such example, the honeycombs are flat packed, expanded on site, and then stacked up with vertical circulation in between the four- or five- story stacks of dwellings. Another approach – going back to the idea of flat packing and expanding – was a unit with a scissor-type mechanism that would expand into a three-floor dwelling. In that way you can achieve three times the density that you would with a simple trailer-home.

A "family" of strategies took the approach of providing a kit of parts; a little flat pack such as you might get from IKEA that you can assemble on site, with the idea that residents themselves could participate such that it would be a fast and easy way to achieve emergency dwellings. In another example, simple temporary scaf-

---

ning are changing completely. The solutions are conditional. They are looking at the immediate picture rather than at the large one.

Claudia Bettiol
Yes, I totally agree. The problem of infrastructure should not be underestimated. Has anyone looked at the minimum skills that a citizen should have to live in a post-disaster city? What are the abilities, the skills, that would allow us to survive without functioning infrastructure, such as sewage or drinking water? What are the skills that we are losing, like being able to work with our hands? What are the skills we have to recover? I see a problem with our lifestyle. Does our lifestyle provide adequate sensibilities to the basics of climate change? Our understanding must

Does our lifestyle provide adequate sensibilities to the basics of climate change?

Global Coalition for Provisional Housing

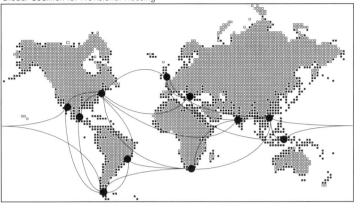

:::: Worldwide flood related disaster risk zones. Inland expansion due to increased hurricane incidence
     compounded by global mean sea level rise. (United Nations Inter Governmental Panel on Climate Change)

● Global S.C.A.Fold Distribution Centers

—— Network of distribution among partnering relief organizations

Regional and Global Implications

Hurricane events tracking through New York State
1888 - 1989
NOAA (FEMA HAZUS Database)

Total Metro Area Population: 41,050,000

Boston, MA
5,400,000
New York, NY
19,500,000
Philadelphia, PA
5,800,000
Baltimore, MD
2,600,000
Washington DC
5,900,000
Richmond, VA
1,200,000
Charleston, SC
650,000

Northeast US Cities and Metro Area Population
Impacted by mean global sea level rise and
heightened flood related disaster risk

Optimal Location

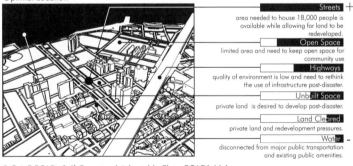

Streets +
area needed to house 18,000 people is
available while allowing for land to be
redeveloped.

Open Space
limited area and need to keep open space for
community use

Highways
quality of environment is low and need to rethink
the use of infrastructure post-disaster.

Unbuilt Space
private land is desired to develop post-disaster.

Land Cleared
private land and redevelopment pressures.

Water −
disconnected from major public transportation
and existing public amenities.

S.C.A.F.FOLD: Self Contained Adaptable Flat - FOLD[able]

The S.C.A.F.FOLD system is a comprehensive provisional housing and recovery strategy. The system uses a
combination of a transforming truck chassis and expandable units to create provisional housing erected
over city streets. The standardization of components facilitates the transportation of units while deployment
over the streets provides access to infrastructure such as utilities and public transportation.

courtesy of Erick Gregory, Jay Lim, Christopher Reynolds

Fig 8.1. What If... Competition Entry S.C.A.Fold

folding, such as you see during typical building construction, would be erected on the vacant site. A kit of parts would infill the scaffolding and people would get their emergency housing very quickly that way. Another category of submissions took a sort of nautical approach. The idea was that New York City is surrounded by water, and there are precedents for habitations on the water such as cruise liners. We even have a floating jail in New York City, which is a floating barge of eight hundred units. So one can envisage a situation where modular container-derived living units are pre-stacked onto barges and stored up the Hudson River. In the event of an emergency they could simply be towed down to the New York City, and people can move in and live in these units on the barge.

Another project looked at a slightly different water-based approach. The idea was that a lot of the debris from the hurricane will be put in the river, and this might create a sort of wet-land situation: pontoons would be put into the harbor and the dwellings would float on these pontoons. The pontoons would provide access to barges that supported the dwelling units. This project is quite reminiscent of the sort of boat settlements that you see in Malaysia.

Many of the entries took a quasi-permanent approach straddling the idea between temporary emergency housing and possibly more permanent. It is well known that it is quite often the case that post-disaster housing becomes at least semi-permanent. In Britain, some of the post World War II housing (known as "pre-fabs") remained occupied and viable 50 or 60 years later. So some competitors asked, "what would it be like if these were here for some time?" They explored designs that created urban neighborhoods rather than simply rows of emergency dwellings.

Many of the competitors addressed the issue of energy supply and

---

evolve relative to the changes we have to try to introduce; and also what we have to teach.

David Burney
The jury was attracted by one of the entries that involved the possibility of "self-help." We would provide the material, and the affected population could, with a little help, participate in building the temporary housing. Jurors felt this possibility could be a very important part of the process, because it gave people an investment in the neighborhood; it gave them participation; it gave them something to do and the feeling to be part of rebuilding the community. Obviously, people who are not physically capable of that would be evacuated. One might question if that is a realistic scenario and if the emergency housing

The jury was attracted by one of the entries that involved the possibility of "self-help." ...because it gave people an investment in the neighborhood; it gave them participation

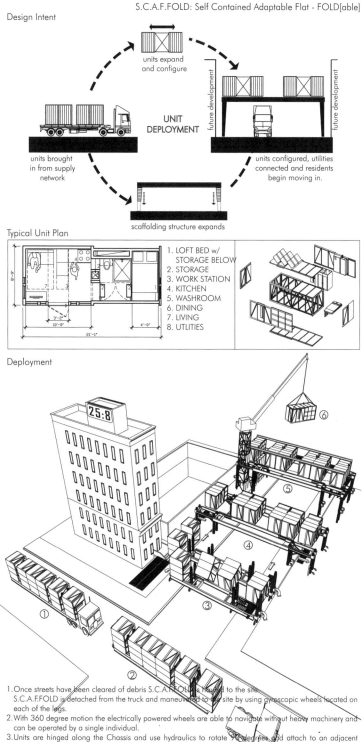

S.C.A.F.FOLD: Self Contained Adaptable Flat - FOLD[able]

Design Intent

UNIT DEPLOYMENT

units expand and configure

units brought in from supply network

future development

future development

units configured, utilities connected and residents begin moving in.

scaffolding structure expands

Typical Unit Plan

1. LOFT BED w/ STORAGE BELOW
2. STORAGE
3. WORK STATION
4. KITCHEN
5. WASHROOM
6. DINING
7. LIVING
8. UTLITIES

Deployment

1. Once streets have been cleared of debris S.C.A.F.FOLD is hauled to the site. S.C.A.F.FOLD is detached from the truck and maneuvered to the site by using gyroscopic wheels located on each of the legs.
2. With 360 degree motion the electrically powered wheels are able to navigate without heavy machinery and can be operated by a single individual.
3. Units are hinged along the Chassis and use hydraulics to rotate 90 degrees and attach to an adjacent Chassis.
4. Base units are secured. Chassis begins to elevate and slide open.
5. Structural beam is installed along the tops of the units to create a truss system.
6. Additional units can be attached using portable cranes. Upon completion, rotate 180 degrees assist with construction on adjacent building sites.

courtesy of Erick Gregory, Jay Lim, Christopher Reynolds

Fig 8.2. What If... Competition Entry S.C.A.Fold

waste management, on the basis that much of the infrastructure may not be viable at the time the emergency housing is needed. Competitors adapted the recent advances in sustainable technology – photo-voltaic cells, wind generators, water treatment, waste treatment, etc, to good effect.

Ten selected projects were developed to a next stage in which some of the technical issues were explored and further developed. Ultimately, the City will adopt some of these strategies to actually incorporate in the emergency flood plan. To be as prepared as we can be, it may be necessary to prefabricate and store hundreds of dwelling units so that they can be mobilized quickly. Several of the competition design approaches seem logical and practical for deployment in the near future. In this regard, two important ideas arose from the competition, apart from the obvious issues of the technical challenge of rapid deployment of emergency shelter. The first is the idea of stacking multiple units to achieve density, and/ or the use of water-borne solutions where available land is limited, seems important if density is to be maintained and wholesale population displacement is to be avoided. The second is the idea that emergency housing may be more permanent that we expect, given the time it takes to reconstruct damaged neighborhoods (as we have seen in New Orleans). The design of the agglomeration of the dwellings therefore becomes key: the circulation and access, the spaces between buildings and the recreational spaces created by the layouts, all help determine whether or not these "emergency neighborhoods" will be places where people will want to live.

1. The competition jurors were: David Burney, Commissioner, NYC Department of Design and Construction; Joseph Bruno, Commissioner, NYC Office of Emergency Management; Paul Freitag, Jonathan Rose Companies; Mary Miss, Artist; Guy Nordenson, Structural Engineer; Enrique Norten, Architect; Professor Richard Plunz, Columbia University.

---

can be sufficiently low-tech to enable relatively unskilled workers to assemble it. That might not be feasible at the higher densities, with multi-story structures.

Antonio Cianciullo
I would like to know how much the City of New York has invested in planning for this scenario. I can imagine that there might be ten prototype structures involved. The first figure would be for the immediate trial testing, and the second figure for the cost of putting the idea into practice. If this stage were reached, what would that figure be?

David Burney
The competition is going to give $10,000 to each of the ten short-listed entries to allow them to develop their de-

signs in more detail. The plan is then to develop and test one or more prototypes so that a fully viable and tested design is produced. When it comes to building the actual structure, and storing pre-disaster housing units, things become more difficult because it would require a substantial amount of money. Because of competing demands on public funds, and because of the uncertainty of disaster predictions, I think it is going to be very difficult to secure the funds that are really needed to prepare in an effective way for the emergency.

Bruna De Marchi
How do you take into consideration the previous lifestyle? Let me give you an example. If it would be the case as some years ago in Italy, in Friuli after the earthquake, I would say that the best choice would definitely be the one that keeps the people near where they lived before. I have no idea what is the best from this point of view in New York, because this consideration depends on lifestyle, traditions, culture, and so on.

David Burney
There has been some history of hurricanes and floods in Europe, before. So, there is some previous experience, and there is a strong emphasis on people staying in their own neighborhood. There is strong emphasis on keeping the population as much as possible where they were living before.

Julie Sze
I am glad that there is finally community planning going on in New York City. My own research on New York focused on environmental justice and community activism under Mayor Giuliani, where there was no planning and there was no sensible environmental policy to speak of. Have we gotten to a post-political state, in the sense that climate change policies will last when political administrations change? Are we in that stage where climate change is not just associated with forward-thinking politicians but it is embedded within an agency, or across agencies, whatever the approach you want to take?

David Burney
I think Julie is right, that every political administration comes and goes and their policy initiatives come and go with them. On the other hand, it is certainly true that now in New York, and it is certainly true for other cities, that things like water supply, energy supply, waste management, are ingrained in the day-to-day business of the city bureaucracy and will not go away with political change. Obviously, some of the planning initiatives that we have

had in the Bloomberg administration may well dissipate, but the core issues on sustainability of cities are permanently with us, and I don't think they will be driven away by politics.

Cynthia Rosenzweig

I feel that climate change is a transformative issue that is creating a new alignment of the different segments of our society. One strand of realignment involves the relationship between practitioners and knowledge providers. It is different, for example, from some of the air pollution work or water pollution work, for which the scientists simply provide information that gets passed on to society with little interaction. I think that the climate change issue is far more transformative. Because climate change knowledge itself is evolving, the interactions between practitioners and knowledge providers need be ongoing. There are many things we still do not know: the sea level rise with the potential for further rapid melting of the ice sheets, or the nature of future hurricanes. Thus, we need to create new modes of working together so that cities will continually have the latest scientific information in workable formats at hand.

CHAPTER 9

Toponymical Rome

Lorenzo Bellicini

If climate change is a global phenomenon, and if the local effects will shift long-standing regional ecological definitions, it follows that old urban political boundaries and their governance structures will have to change as well in order to implement effective policy. This question arises in relation to the number of other indicators that point to the general obsolescence of municipal boundaries relative to the next generation of town-planning policy imperatives. It may be useful to examine this question in relation to Rome and its territory.

In 2001, the Planning Office of the Rome Municipal Administration appointed CRESME (Center for Economic, Sociologic, and Market Research for Regional Planning) to conduct research on a variety of topics, including identity, centrality, and new local municipalities within the City of Rome.[1] Rome encompasses a territory that totals 130,000 hectares, with a municipal area large enough to contain the rest of the metropolitan region within its borders. This Greater City is governed on the basis of historical administrative subdivisions, mainly through *circoscrizioni* (boroughs). It has been proposed that for the purposes of creating new local municipalities, borough functions be expanded. The Planning Office wanted to determine the existing as well as the potential identities that could determine the creation of new municipalities, and so commissioned the CRESME to take on this task.

Among the results of the sample interviews with Rome residents, a major theme, suitable as a model of interpretation for the whole

---

Richard Plunz

We must question the relevance of our institutions, in general, and especially the institutions of the Western world. We have our ways of doing things but we have to be honest with ourselves. Our modes of strategic thinking can fail us, in the face of the phenomenal growth of Lagos, or even of Caracas, where you have a region of five millions inhabitants and 60-70% of illegitimate buildings. Who says what is "illegitimate?" Maybe illegitimate is actually legitimate in that context. What are our tools in this context? What are our tools institutionally and otherwise?

Who says what is "illegitimate?" Maybe illegitimate is actually legitimate in that context

Matteo Caroli

I would like to remind you of the case of several large

research project, quickly emerged. A strongly-rooted resident population with powerful micro-level identity leanings characterize the urban environment within Rome. When residents were asked whether they wanted to relocate to another part of the city or to another city altogether, 80% indicated their desire to remain in their neighborhood. Even if they wanted to move, it would be within the confines of their neighborhood. Because Rome is so large and composed of many different segments, familiarity with a certain place and the resulting confidence has become an urban value. Paradoxically, this is independent from the commercial value of the place. The principle of strong residential identity in the population applies to both the central and the outlying boroughs, encompassing areas with both numerous as well as few functions.

Residential identity and centrality do not necessarily coincide. This discovery paved the way for another model of understanding and gave rise to new questions: What are the borders of the urban areas as defined by the principle of residential identity? And furthermore, what are the exact features defining these areas? This required a new line of research involving the drawing-up of different urban maps, with the gradual creation of a synthesized and verifiable system by which field interviews were conducted with residents who were considered the primary "experts on the area." They would be able to sufficiently identify the toponyms and borders of Rome's "micro-cities." The result would yield a map of 200 micro-cities comprising Rome. The final stage of this process was to verify the field interview results with the administrators of the 19 boroughs.

This work could empirically prove the common notion that Rome is a city composed of a galaxy of micro-cities. Each micro-city defines its own toponyms and borders, which in turn become the fountainhead for re-interpreting the metropolis. More specifically,

---

UK cities, Manchester, Liverpool, Shieffield, in addition to London, of course. They have been at the heart and soul of the British industrial revolution. In the '60s they went through a major crisis because of the decline of the steel, coal and shipyards industries; these cities seemed to be dead, they were the Lagos of Europe. Now these cities have become very competitive, very attractive. They have been able to identify and to implement a policy for their local areas where they combine urban planning, landscape, and social aspects. They overcame their problems and became a symbol of excellence, both in Europe and in the UK.

Domenico Cecchini

If you start at the community level, you can generate this

If you start at the community level, you can generate this creative participation, which gives rise to innovation, in the solutions too

each micro-city would have a reconstruction of the system of functions. The "centralities" would be highlighted and basic statistical information would be provided on the population, households, housing, and the economy. The Toponymical Map of the micro-cities would become an alternate system of knowledge about the Rome area, and a tool for understanding and participating in planning contingencies seen and unseen, including of course, various climate change scenarios.

Toponyms and Identities, Information Units Endowing Meaning

Based on the sampling of resident populations, Rome is a composition of a series of individual urban regions with strong residential identities. These regions have their own borders and names. The Rome Toponymical Map becomes the setting in which the historical and the more recent features of modernity are juxtaposed. This can give rise to a deeper interpretation. Within the territory of Rome, there are regions that are less known, more distant, and more scattered. This diversity has been a recent development. These places represent the city's unknown or untapped identities and convey a limitation deriving from its large size. Rome is a complex, multifaceted city and most of its citizens make a "limited" use of its urban space. This use is related to a few places, and the city is perceived by "points" or "micro-areas" as well as by crossing routes.

In the best of cases, Rome's residents are familiar with three, four or even five micro-cities: the place where they live, the place where they work; the city centre and/or the place where they go to enjoy themselves or for shopping; and not much else. The city is therefore experienced through the known micro-cities and the "routes" connecting the micro-cities. In other words, since people are aware of local happenings, they perceive measures affecting the city and city

---

creative participation, which gives rise to innovation, in the solutions too.

Matthew Nisbet
At the local city scale there is really a lot of potential for social marketing campaigns, motivating individuals to adopt personal behaviors leading to energy conservation or public transportation. Current social marketing campaigns have very little to do with informing the public about the state of the science or increasing their knowledge. Instead, they deal with framing; putting the issue in the context of their social identity or something they already care about; or framing the issue in terms of social norms - what everyone else is doing, or shaping the perception of majority behavior.

Current social marketing campaigns have very little to do with informing the public about the state of the science or increasing their knowledge

routes according to their own inhabited spaces and perspectives. The Toponymical Map with individual residential identities takes on meaning due to the results achieved by the detailed analysis and due to the consequent interpretation of current processes (hierarchies of micro-cities, functional specialization, urban dynamics, economic and social characteristics, town-planning standards, userships, etc.).

The Map serves as a potential contributor for comprehension and management of Rome's complexity, augmenting other urban models in use. For example, the National Statistics Institute (ISTAT) and the Rome Municipal Statistics Office's methods of dividing the city into over 17,000 census sections. Within this array, the micro-cities are meaningful local units, recognizable by the residents who associate a name with an area. Consequently, the map of the micro-cities becomes a useful way of aggregating information on the basis of meaningful local units comprehensible by people who live throughout the city. The existing way of understanding is through the census framework that is neither recognizable nor qualitatively defined, as it is only a number indicating population. Initially, it was an attempt to create an information system based on local units recognizable by its residents, and designed for updating. The goal was to formulate a system for the reduction and synthesis of Rome's urban complexity while serving as a major tool for participation based on understanding.

Recognition of Identity and Consensus
The creation of the Toponymical Map, which is recognizable by the population, has served as a highly regarded and verifiable means of communication and transparency. For example, "Dragoncello" and "Dragona," though adjacent urban areas, are indeed very different, and the resident population recognizes the paradigm of dif-

---

Maria Paola Sutto
An aspect that is missing completely from this debate is the education of younger generations about climate change. We are bequeathing, bestowing on them a very heavy legacy, without educating them. There are not enough education projects, at least in the United States, to teach them to live in a sustainable way.

Cinzia Abbate
As an educator being involved with the Rensselaer Polytechnic Institute, and having followed education policy also within the European Community I would say that in the last ten years a lot of work has been done through the universities to really bring the issue of ecology to the

An aspect that is missing completely from this debate is the education of younger generations about climate change

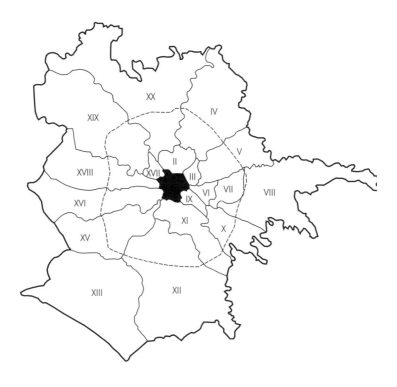

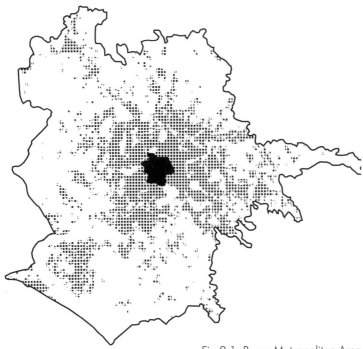

Fig 9.1. Rome Metropolitan Area
Administrative Boundaries (above), Urbanized Areas (below)

ferent identities. The analysis of micro-cities reinforces an identity-oriented policy. By open dialogue and discussions, the Map-making process served as a substantial means for residents, local groups, and local institutions to participate in planning decisions. The attributes and identities of the micro-cities, typically lost in Rome's complexity, become the basis for the city plan's analytical strategy. They serve as the initial factors of recognition to be considered in a dialogue. The size of the micro-cities enables us to handle the complex planning procedure in two directions: primarily by recognizing the demands of the individual micro-cities; secondly, through a process of aggregating the micro-cities by *circoscrizioni* or functional sectors and thematic areas for intervention. This enables definition of the larger scale urban strategies that ensure the inclusion and integrity of the micro-city system.

Complexity, Lack of Knowledge, Routes

The need for a micro-cities map highlights the lack of knowledge about Rome among its residents. Once again, because Rome is a complex city, its residents make a limited use of the urban space and typically stay within the confines of what is familiar. There are people who have never been to visit the "Rome" understood as the city centre. Due to the complexity and unfamiliarity of a place, a visit to or an encounter with another micro-city can make people feel uncomfortable and wish to return to their known habitat. This attitude explains the strong sense of residential roots even in underprivileged conditions.

The city is, therefore, perceived through micro-areas, which refer to the known areas and through "routes." The theme of routes is a fertile one with regard to both the routes within the micro-areas and the more traditional "routes" between the micro-areas. The routes between the areas are the links conditioning mobility, but

---

forefront. Is this really effective yet? I don't know, but, for example, we really get an impressive number of students who constantly call us in our studio asking for information about projects that we have done. So, there is a growing interest that comes from the university.

Claudia Bettiol
I can give another example. At University La Sapienza, in Rome, a new policy is being implemented whereby in order to get their degree, students will be required to substantially study environmental and energy sustainability. This decision will touch 50,000 students every year, whose degree areas will be as diverse as humanities, veterinary sciences, and pharmacy. On the University website, students have become aware of the fact that they

There is a growing interest that comes from the university

they are also "routes of meaning" that can orient people's behaviour. When asked whether residents took part in neighbourhood life, the popular response was "absolutely not" (40% of the interviewees), and the research also shows that the main problem is a lack of meeting places.

Only 16.2% of the residents said they "fully take part in neighborhood life," while 33.1% stated that they "take part enough." Another 11% said they take very little part in the life of their neighborhood. Therefore, for over 50% of the population, a strong residential identity is accompanied by a low usage of their urban residential environment. The main model of the use of urban space appears to be home-route-home (home-route-work). Residential roots depend on the knowledge accumulated with respect to confidence and trust, more so than on the residential area's value. The poor use of urban space indicated in this model exemplifies the indifference to issues such as centrality and local roots, even in zones without many functions. This aspect also builds upon the fundamental role between areas and routes in the urban space context along with mobility themes.

Micro and Macro Levels in Town-Planning Policy
In light of the previous considerations, guidelines for Rome's town-planning policy can be delineated. Primarily, there is a need to respond to residents' preferences, as well as to investigate the specific aspects of the micro-cities with which the residents identify themselves; this would serve to define a framework for specific interventions. If the goal of urban policy is to improve the living conditions of its residents, the micro-city approach is one of the first levels that must be adhered to in the town-planning strategy. Town-planning policy will only become successful if it targets interventions in the micro-cities.

---

have to fulfill this mandatory requirement beginning in September 2008, but the University's phone call operators have been overwhelmed from students who wanted to take that class now. They don't want to graduate without having had this opportunity.

An organic plan for micro-interventions therefore has a surprising potential for consensus, particularly if it is based on the central themes of "functional routes," where structuring and linking of various existing and/or potential functions take place. These interventions involve a complex variety, ranging from mobility to the functional revision of individual urban spaces. Naturally, this policy of systematic micro-intervention in the micro-cities must be superimposed on a larger scale strategic policy, within the context of individual boroughs, as well as on the metropolitan scale. Our research shows that the interpretation of the micro-cities' functional characteristics - their differences and specializations - helps us to deliberately create a plan with various levels of intervention based on the themes of centrality. These include measures for: the consolidation of the central systems on the local level, calling for the identification of new public areas as well as the integration of existing functions; the reorganization and integration of the central systems on the borough level by identifying how the location of the system relates to its main structural routes; the identification of the relationships with the adjacent systems (new routes and directions of development); the redefinition of the areas on the edge of the system (hypotheses for new uses, installations, parks, etc.); the redefinition of the elements and components of the central system on the urban and borough level, and the opening of the relationship system to the surrounding areas; and the reorganization of the elements and the functional and structural components of the specialized centralities on the metropolitan level.

1. CRESME, *Micro-cities: A New Toponymical Map of Rome*. Study for the 2003 Master Plan of the City of Rome, Rome, 2001.

CHAPTER 10:   **Communicating the Safe City**
Antonio Cianciullo

"La Città Sicura," or "The Safe City," is the Italian title of a science fiction short story written by Andrew J. Offutt in the early 1970s.[1] In a world dominated by the fear of pollution, the appearance of an old, illegal, petrol-driven automobile on an urban freeway triggers a reaction of rejection and a spiral of violence that climaxes with the security forces dropping bombs from a helicopter. Thirty years ago, only a science fiction writer could have imagined a world in which environmental concerns once again unleash violent atavistic behavior. In November 2007, it was the illustrious Center for Strategic and International Studies in Washington that produced a study entitled *The Age of Consequences*, showing how climate change and the associated rise in sea level are physically endangering the safety of many cities and the level of freedom that we now take for granted.[2]

From the scientific viewpoint, the leap was accomplished with the fourth report of the International Panel on Climate Change (IPCC). For all the uncertainty remaining in many fields, the enormous amount of work still to be done, and the countless problems to be solved, the scale of the challenge is now clear to the scientists and to all of us gathered together here in this room. And yet, when we leave here and go to a restaurant, to the hotel, or back to our homes, the options facing us will be at variance with what we have just discussed. It is not as if as individuals we can just go on a diet if we want and reduce our intake of fat. This approach will hardly produce a significant decrease in carbon emissions.

Giuseppe Tripaldi
Regarding the information now available on climate change, there seems to be a kind of communication fog. It's always very difficult to find your way when you are in a fog, and when engaged in a decision-making process, you would like to have some well-defined alternatives in front of you. We need to arrive at a simplification or reduction of the great mass of information available.

We need to arrive at a simplification or reduction of the great mass of information available

Antonio Navarra
We are surrounded by words and images and sounds, but I am not sure that they correspond to real information. There are a lot of words, but often weak content. The fog that is cloaking climate change is much thinner than we think, but certainly we need the capacity to separate the

The variance between science and the wider public, which is at least partially inevitable, shows that intellectual awareness of the climate problem has not yet become common knowledge. Research has yet to transform its potential into objects of low environmental impact. Cities are still stuck in outmoded forms of organizational structure. The day-to-day services provided in terms of mobility and building temperature regulation are still based on a scale of values in which energy is not a shared asset to be administered with care, but plunder to be wrested from hostile nature. What stops us from making the leap from the cultural universe of fossil fuels to a world of efficiency and renewable energy? A great many things: the interests of economic groups based on the old structures; the time physiologically necessary to move from the industrial age to a new era; and the complexity of this transition. But there is also a failure in communications that has only started to improve in recent months.

Despite the steady increase in the space devoted to environmental questions by the media, there is still an imbalance in certain areas. What is it that frightens the half of the world population that lives in cities and will soon become the absolute majority? Let's take the fears of the inhabitants of Italian cities as an example. If we adopt as our yardstick of mood the proposals of a political class whose slogans are increasingly coined in accordance with the "just-in-time" management philosophy, we can see that attention is largely focused on announcements of sensational measures to fight crime. The fear of falling victim to a rising tide of criminal acts appears to predominate in Italy, as well as other countries where the situation is similar in some respects. And stress is laid in particular on the dangers arising from the ever-larger numbers of immigrants.

What figures exist to justify this fear? In 2005, the most recent year

---

fluff from the real stuff. What is important, for instance, is not only what has been said, but also who is saying it. The issue of reliability of the sources becomes incredibly important, because it is so easy to send words to anybody nowadays. We are truly in an information world. It used to be difficult to find an answer to a question, but now you can find a thousand answers to a question on the internet. So the issue now is not to find the answers, but to detect which answers are true and reliable. Professionals of the information industry, of the media industry, meet these problems every day, but they are also trickling into the scientific world with the multiplication of journals and paper archives.

The issue of reliability of the sources becomes incredibly important, because it is so easy to send words to anybody nowadays

for which official statistics are available, there were 710 murders in Italy (with a drop to 621 in 2006); 4,000 cases of sexual violence; and 46,000 robberies. Statistically speaking, the odds against being murdered in the course of a year in Italy are just under one hundred thousand to one. You are four times more likely to die on the field playing American football. Let us compare these figures with the possibility of loss of life due to an illness deriving from the pollution of our cities. According to a study published by the World Health Organization in 2006, 8,220 deaths could be avoided in 13 Italian cities by achieving the target set by Brussels of bringing the volume of fine particles down to 20 micrograms per cubic meter a year by 2010.[3] Given that traffic is responsible for about 70 percent of the particulates in the urban areas in question, it can be stated that the chances of dying in the 13 major Italian cities because of the present system of urban transport are 12 times as high as those of being murdered.

In short, we are scared of the masked killer lurking in waiting around the first dark corner, yet we walk fearlessly in the midst of exhaust fumes that are 12 times more lethal than this figment of our imagination. This contradiction is confirmed by a report on crime and safety in Europe funded by the European Commission and released in Brussels in February 2007.[4] According to this study, while the number of people with direct experience of crime is decreasing, there is a growing fear of being attacked and mugged at knifepoint. Some 30 percent of the European Union's citizens are afraid of being robbed and do not feel safe in the street, even though the same people recognize a marked drop in crime. So there is a widening gap between risk and the perception of risk. We are more afraid of the sudden attack than the slow, insidious threat. As primates, we have by nature a deep-seated terror of being suddenly pounced upon by a feline predator. And this continues to dominate

Fig 10.1. Odds to Die and Climate Change

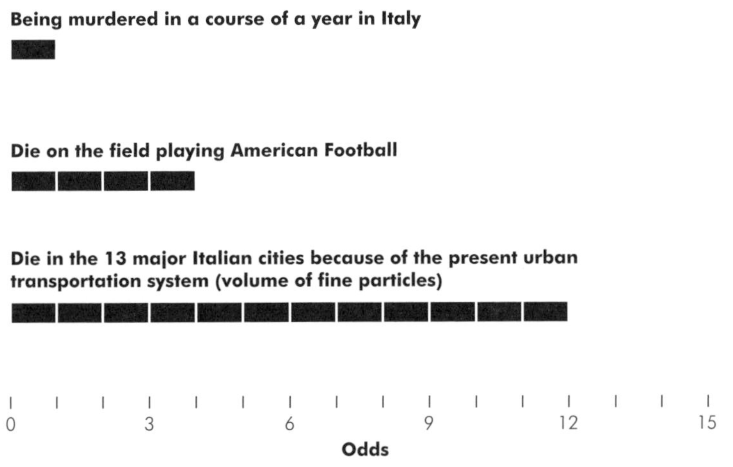

Being murdered in a course of a year in Italy

Die on the field playing American Football

Die in the 13 major Italian cities because of the present urban transportation system (volume of fine particles)

us even when lions and tigers have become species in danger of extinction and our greatest danger is of following in their footsteps, dragged down by destruction of the climatic balances that have accompanied us for thousands of years, albeit with some significant swings.

It will be the task of physicists, biologists, chemists, and economists to point out the gradual evolution of risk and the impacts on health, the economy, nature, and art connected with this slow, insidious shift in climate. Communicators must, however, immediately address the political nature of the failure to deliver the message and the consequent stifling of alarm about the effects that climate change will have on our cities. And from this point of view, the ever-closer interweaving of politics and information appears to be anything but a coincidence. This is a combination that manifests itself most clearly in the case of Al Gore. Defeated as a politician, he has become a winner by emerging as a worldwide multimedia champion in the battle against climate change.

We can, therefore, look to Al Gore's latest work for an initial indication of the course to be taken if we are to convey the message of a radical change in environmental behavior as the prerequisite for revitalization of our cities. As he points out, President Bush could never have called a law increasing air pollution the "Clean Skies Initiative" or a law allowing the felling of trees in national forests the "Healthy Forests Initiative" if he had not been sure that the public would never find out what the details of those laws envisage.

The first critical area that undermines the credibility and efficiency of the information system is, with the due exceptions, the excessive sensationalism and distortion of words to generate a form of communications that is increasingly divorced from reality - a lot

---

Cynthia Rosenzweig

The IPCC process entails scientists coming together with government representatives. First there was work in teams with developed and developing country scientists. It takes four years of actually working together to produce a massive report. So even if the results are contentious, I feel that there was a new way that we began to work together in the world. And what do you do as a scientist in this process? You bring the report, and the Government representatives are there. There is disagreement. A small group of some of the delegates who don't agree, and scientists who don't agree are assigned to a contact group. And someone is made the leader. So, in one case, a young negotiator from Saudi Arabia was the leader. He was very good at working with the disagreement in the

It takes four years of actually working together to produce a massive report

of cream and little substance - more marketing than news. This is the subject addressed by the highly successful Italian journalist and opinion-maker, Marco Travaglio, in his latest book, the title of which translates as "The disappearance of facts: please abolish news so as not to disturb opinion."

If we all gather together in a soundproof room for a few days and talk about the weather, barring the doors and windows, pulling down the shutters and keeping our cell phones, radios, televisions and computers all switched off, everyone will be entitled to maintain that it is sunny outside or pouring down with rain or hail or snow. Each opinion will have the same dignity, credibility, and reliability. And the discussion can go on with great intensity for days or weeks. Or at least until one of the inmates suddenly decides to open a door or window, to phone someone outside, to switch on the TV or radio, or to get on the Internet to find out what the weather is actually like. Assume that he or she finds out that it is raining. Everyone will have to recognize this and from that precise moment on nobody will be able to claim that there is sunshine or snow or hail. The debate is over and we can all go home. But if things can be arranged so that none of the inmates can make contact with the outside, the weather will remain the subject of debate forever with opposing views that, in the absence of any hard facts with which to compare them, will maintain all their validity intact.[5]

If we stop at this initial level, however, it might be imagined that activating the classic antennae of the information system will be sufficient to bring out the concealed evidence of the danger of climate change: more news, more in-depth coverage, more interviews with scientists. But this is precisely what has happened over the last few years, and the results are no more than partially satisfactory.

---

room. He would say, "Morocco, what do you think? And "Tunisia, are you coming on board?" And this feeling came over me, in that room, that there we were with tremendous contention from different, truly different cultures and points of view on the issue, but we were working on the future of the planet, and not having disagreement dissolve into war. So, I really do believe in the possibility of transformation through addressing the climate change challenge.

I really do believe in the possibility of transformation through addressing the climate change challenge

Matthew Nisbet

There is a lot of communication about the science of climate change, and there will continue to be a lot of communication about the science. The unfortunate thing is that this dimension is really only followed by people

It is therefore necessary to identify other levels of problems in the field of communications. A second level is connected with the difficulty in getting oriented within the growing sea of information. According to calculations carried out by the University of California at Berkeley, the information produced between 1970 and 2000 was equal to the amount previously accumulated ever since prehistoric times. The same volume of information was again produced in the next three years, and the information produced in 2020 will be 50 times as much as in 2003.[6] The series of inadequately filtered news items, the succession of contradictory echoes, and the phony scoops arrogantly splashed across the front pages all generate disorientation. And this disorientation combines with the over-accumulation of information to generate a form of "Stendhal Syndrome:" panic induced by the impossibility of keeping up with the rising tide of news.

Given these figures, we can understand why Richard Saul Wurman stated in *Information Anxiety* that the more we are bombarded with images, the more distorted our vision of the world will be.[7] The more time we devote to the reports on individual events, the less time we will have left to grasp the underlying whys and wherefores, to comprehend structures and relations, and to understand the present in relation to historical context. We are hypnotized by a flood of superficial fact, rendered mute, passive, and impervious by a surfeit of data that we have neither time nor resources to transform into valuable information. While the dangers described here naturally regard the entire spectrum of communications, information on climate deserves particular attention because it is no easy matter to describe a chaotic system, and because there are particularly strong interests seeking to make this task still more difficult. To cite just one example, in September 2006 the Royal Society accused Exxon of having paid anti-environmentalist lobbies $2.9 million to mini-

---

who already are interested and attentive to the science of climate change.

They are also people who are already concerned. When they hear about the emerging science, they become even more concerned, whereas people who are skeptical are people who are not even focused enough to care about the science, or automatically reject the science based on ideology.

Lieven De Cauter
If you call someone "alarmist," it is like saying "conspiracy theory." You neutralize his voice. When you say alarmist to someone, you just tell him he is a leftist. Of course you all know this. So I think we should rehabilitate alarmism.

I think we should rehabilitate alarmism. We should find a word for the counter-strategy: "quietist," or "liar," or whatever, but a good word

mize the risks connected with climate change.

Finally, there is the third level of problems connected with risk perception. Our selection of alarms is in itself useful because it helps us to survive. If we took all the unlikely disasters into consideration, we would not only develop ulcers but also end up with no energy left to focus on the real problems. The unconscious can, however, betray us. An overly-alarmist piece of information can trigger a dangerous form of protective self-censorship, as demonstrated by a survey carried out in a community living beneath a dam. The level of fear was found to increase together with proximity to the dam, but registered a sudden fall in its immediate vicinity, since the inhabitants of those houses were psychologically incapable of coping with such a degree of anxiety and suppressed their awareness of the problem. This is why some environmentalist associations have reshaped their communications strategy in recent years to strengthen the message of hope connected with possible change.

These are the problems, and they are by no means negligible. There are also, however, many good reasons for a qualitative leap forward in information on the climate. The serious consequences of climate change are becoming increasingly obvious. The pressure of public opinion is mounting and effecting a change in the attitudes of the strongest countries. And economic interest in a model of production with lower environmental impact is gathering momentum.

What is needed at this point is a calibrated message of alarm in which attention can start to focus on what can effectively be done to defend and perhaps even improve the role of cities in the era of climate change. What adjustments will be needed to the rules of urban planning? What forms should building take? What will be the best type of transport? Who is to pay for this change? Who will

---

We should find a word for the counter-strategy: "quietist," or "liar," or whatever, but a good word.

If humanity does not shift into emergency mode, we are really not only damaging humanity or the human race, but even more importantly, our planet for hundreds if not for thousands of years. So, this is my plea, and it will be a continuous plea - for alarmism and for re-habilitating alarmism. For me it is very clear. If we don't get to very concrete actions that alert people, that change peoples' habits very concretely, we may just as well lay back and do nothing.

benefit? Who will be out of pocket? These are the concrete questions to which answers must be found.

The much-discussed legacy that we have received and must hand on to our children's children is not confined to "nature." Staying in the vicinity of this venue also includes what Ferdinando Gregovius spoke of in his description of his journey in Italy: "ancient towns, mostly episcopal sees, perched on green hills of chestnut trees and olive groves or garlanded with vines and giving the countryside of the Lazio region a predominantly historical imprint."[8] We must try to preserve what remains of these towns and landscapes.

1. A.J. Offutt, "Meanwhile, We Eliminate," in Roger Elwood, ed. *Future City*, (New York: Trident Press, 1973).

2. Center for Strategic and International Studies, *The Age of Consequences: The Foreign Policy and National Security Implications of Global Climate Change,* Washington, DC, 2007 http://www.csis.org/component/option,com_csis_pubs/task,view/id,4154/type,1/ (accessed July 2008).

3. World Health Organization, *Health Impact of PM$_{10}$ and Ozone in 13 Italian Cities,* Copenhagen: 2006. http://www.euro.who.int/document/e88700.pdf (accessed July 2008).

4. European Commission 2007, *The Burden of Crime in the EU: A Comparative Analysis of the European Crime and Safety Survey (EU ICS) 2005,* http://www.crimereduction.homeoffice.gov.uk/statistics/statistics060.htm (accessed July 2008).

5. M. Travaglio, *The Disappearance of Facts* (Milan: Il Saggiatore, 2006).

6. P. Lyman and H.R. Varian, *How Much Information?* (University of California, Berkeley, CA, 2003). http://www2.sims.berkeley.edu/research/projects/how-much-info-2003/ (accessed July 2008).

7. R.S. Wurman, *Information Anxiety* (New York: Doubleday, 1989).

8. F. Gregovius, *Passeggiate per l'Italia* (Rome: Avanzini e Torraca, 1968).

CHAPTER 11:   **Real People, Urban Places**
Matthew C. Nisbet

Many observers point to 2007 as a major breakthrough for engaging the American public on climate change. Yet despite record media attention and the strongest conclusions to date by the scientific community, survey data show that Americans still remain uncertain about whether climate change is a problem, and whether or not it deserves to be a political priority. Public opinion obviously matters. At the national level, as long as climate change remains a non-issue for the American public, it will be very difficult for elected officials to reach a consensus on major policy action, and for the United States to participate in international agreements. Policy gridlock, therefore, is in part a communication problem.

Solving the public opinion challenge means defining the complexities of climate change in a way that connects to the specific core values of a diversity of citizens. Not every citizen defers to science or cares about the environment, yet among climate change advocates, these points of reference continue to be the dominant communication emphasis. As a result, audiences already concerned about the problem grow more intense in their beliefs, while many Americans literally tune out the message. To fix this problem, a fundamental communication shift is needed. Climate change needs to be repackaged around core ideas and values that a majority of Americans already care about. This means shifting the public lens away from distant artic regions, socially remote people and places, or consequences far off in the future. Instead climate change must be recast as an urban problem with local impacts and solutions. While there are factors unique to the United States context, similar

---

Claudia Bettiol

Three years ago for the very first time, the regional governments in Italy devised a strategic plan to develop renewable energies. Part of this plan included a section dedicated to communication, to promote shared knowledge. We started from the assumption of a sustainable transformation; a transformation that managers would term both as a product and a process transformation. It followed the logic that if I have to revolutionize the city, **If I have to** first I must have a product - perhaps an environmentally **revolutionize** friendly house. Then when we extend this to the urban **the city, first** scale, through replication, we are processing innova- **I must have** tion. When a couple of years ago we started to outline **a product -** a strategic plan for the Lazio region, we focused on the **perhaps an** fact that it would not be only architects or engineers who **envirnmentally** **friendly house**

communication principles apply across other countries.

The Origins of Perceptual Gridlock

By the end of 2007, the publicity and box office success of Al Gore's documentary, *An Inconvenient Truth*, had been backed up by ever stronger expert consensus from the Intergovernmental Panel on Climate Change (IPCC). Yet despite all of the media attention generated by Gore and the IPCC, climate change still rested relatively modestly on the overall news agenda. In fact, according to data tracked by the Pew Project for Excellence in Journalism, climate change failed to crack the top twenty most covered news stories of the year.[1] Moreover, despite the strongest conclusions to date by the scientific community, the American public still remained uncertain about whether the majority of experts agreed on the matter. Depending on how the question was asked, belief that scientists had reached a consensus view ranged from only a third of Americans to more than 60 percent. This variability reveals a "soft" public judgment that continues to be susceptible to the misleading counter-claims of many political conservatives.

Views on expert agreement are not the only areas where public opinion remains tentative. When asked in comparison to other issues, global warming scored consistently as a low political priority. And in open ended questions asking Americans to name the most important problem facing the nation, global warming registered routinely at less than 1% of responses. Public judgments of the objective reality of global warming also vary widely, forming what I call a "Two Americas" of climate change perceptions. The divide starts at the top. In a 2007 *National Journal* survey of members of Congress, a mere 13% of Republican members said they believed that the earth was warming because of man-made problems, compared to 95% of their Democratic colleagues.[2] The public breaks

---

were considered experts in the specific segments, but we needed to involve knowledge that was shared by several disciplines.

Matthew Nisbet
You can reframe the climate issue, in the process triggering attention from other policy contexts such as the Centre for Disease Control or the Pentagon. This raises and elevates the issue on the policy agenda, but you are also casting it in ways that make it either of interest or personally meaningful to an audience group that currently doesn't really pay attention to it.

Harriett Bulkeley
All of the variations on frames still try to communicate to

We needed to involve knowledge that was shared by several disciplines

down along similar party lines. Gallup found that between 2006 and 2007, worry about global warming grew to a record high of 85% among Democrats, while the percentage of worried Republicans remained unchanged at 46%.[3] When you factor in education, an even deeper chasm is revealed. According to a Pew survey, only 23% of college-educated Republicans said that global warming was due to human activity compared to 75% of their Democratic counterparts.[4]

So at the end of 2007, despite Gore's breakthrough success with *An Inconvenient Truth*, American opinion was little different from when the film premiered in May 2006. Gore and the spike in mainstream media attention had intensified the beliefs of Americans who were already concerned about climate change, but a deep perceptual divide between partisans remained. As editor Donald Kennedy lamented in his end of the year editorial in *Science*, this continued climate gridlock rated as the "science breakdown" of the year.[5]

What then explains the difference between the objective reality of climate change and its perceived subjective conditions? If science and mainstream news attention alone drove public responses, we would expect increasing public confidence in the validity of the science, and decreasing political gridlock. However, instead of scientific reality, ideologically driven interpretations are providing the dominant perceptual cues for the public. Although the George W. Bush administration now accepts that human-induced climate change is real and that action is needed, several conservative think tanks, political leaders, and commentators continue to hew closely to the decades-old playbook on how to downplay the urgency of the issue. Moreover, even as Republican leaders such as John McCain and Arnold Schwarzenegger assert the need for action on global warming, the strength of these decades-old oppositional

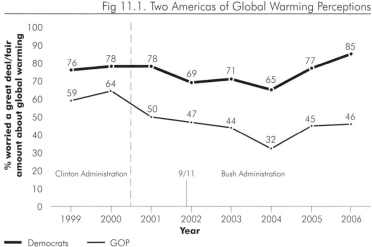

Fig 11.1. Two Americas of Global Warming Perceptions

Democrats —— GOP

Source: Nisbet & Myers, *Public Opinion Quarterly*, 2007

frames remain salient in popular culture, political discourse, and the memory store of many citizens.

During the late 1990s, the climate-skeptic playbook was in part devised by Republican pollster Frank Luntz. Based on dial groups and polling, Luntz recommended that the issue be framed narrowly in terms of scientific uncertainty and economic competitiveness. This "paralysis by analysis" strategy was effectively implemented by conservative think tanks and members of Congress to defeat adoption of the Kyoto treaty and other major policy proposals. The strategy also led to distortions in news coverage. As political reporters applied their preferred conflict frame to the policy debate, they engaged in a "dueling experts" style of false balance that added further distortion to the science.

Senator James Inhofe (Republican-Oklahoma), former chair of the Environment and Public Works Committee, remains the most visible voice of climate skepticism. In speeches, press releases, and on his Senate blog, Inhofe casts doubt about the overwhelming consensus view on climate change, selectively citing scientific-sounding evidence. To amplify his message, Inhofe takes advantage of the fragmented news media, with appearances at conservative talk outlets such as Fox News and Web traffic driven to his blog by The Drudge Report. For example, in a February 2007 *Fox & Friends* segment titled "Weather Wars," Inhofe deceptively argued that warming was in fact due to natural causes and that mainstream science was coming around to this conclusion. As Inhofe asserted, unchallenged by the host despite the reality of the science, "Hollywood liberals and people on the far left such as the United Nations" want the public to believe that global warming is man-made. Similar storylines of scientific uncertainty and damaging economic impacts continue to be pushed by other conservative commenta-

---

individuals. They are trying to change the attitude and behaviors of an individual. Bruna De Marchi's talk raises the notion of a more collective response; issues about trust and how the people work together. In the U.K. context we have over a decade of research, mainly by sociologists, showing that the gap between the public perceptions of climate change and public action relates really to the issues of responsibility and trust.

Julie Sze
If terrorism and global insecurity are issues that touch Republicans, Democrats, Independents, why not reframe climate change as a global instability?

The gap between the public perceptions of climate change and public action relates really to the issues of responsibility and trust

tors, including influential syndicated columnists George Will and Tony Blankley.

Al Gore, many environmentalists, and even some scientists have attempted to counter the uncertainty and economic development frames with their own Pandora's box emphasis on a looming "climate crisis." They pair this interpretation with a social progress narrative of modern civilization that was once in harmony with nature, but because of industrialization and consumption, has thrown the Earth into dangerous disequilibrium. To instantly translate their preferred interpretation of "climate crisis," environmentalists have relied on depictions of specific climate impacts including powerful hurricane devastation, polar bears perched precariously on shrinking ice floes, scorched earth from drought, blazing wild fires, or a future where sea level rise has put famous cities or landmarks under water. With an accent on the visual and the dramatic, this strategy has been successful in triggering similarly framed media coverage. For example, a much talked about *Time* magazine cover from 2006 featured the image of a polar bear on melting ice with the tagline: "Be worried, be VERY worried."

Yet this line of communication plays directly into the hands of climate skeptics, only further reinforcing a "Two Americas" of climate change perceptions. As Andrew Revkin of *The New York Times* relates, given that the error bars of uncertainty for each of these climate impacts are much wider than the general link between human activities and global warming, these claims are quickly challenged by critics such as Inhofe as liberal "alarmism," putting the issue quickly back into the mental box of scientific uncertainty and partisanship. These types of environmental fear appeals, especially when lacking specific recommendations for how citizens can respond to the threat, are also likely translate into a sense of fatalism

Fig 11.2. Combined Coverage at New York Times and Washington Post

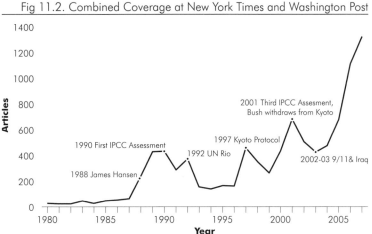

Source: Nisbet & Myers, *Public Opinion Quarterly*, 2007

on the part of the public.

New Mental Boxes and Social Locations

Despite the continued perceptual gridlock, this past year also featured important innovations in climate change communications.
Several interpretations emerged that have the potential to move beyond polarization and to unite public perspectives around common
goals. For the public, a complex issue such as climate change can be
the ultimate ambiguous threat, meaning that depending on how the
problem is "framed" in news coverage, the public will pay more attention to certain dimensions or considerations of global warming
over others. As decades of social science research have concluded,
often the public will choose the interpretation of an issue that is
most consistent with its existing preconceptions or social identity.
These messages then activate a train of thought that leads to very
specific attributions about the nature of an issue - who or what
might be responsible for a perceived problem, and what should be
done in terms of policy.

Activating concern and catalyzing behavior change across key segments of the public depends on establishing the right perceptual
context. The communication challenge is to shift climate change
from the mental box of "uncertain science," an "unfair economic
burden," or a "Pandora's box" of disaster toward a new cognitive
reference point that connects to something the audience already
values or understands. Over the past several years, several strategic interpretations focus attention on adaptation strategies rather
than simply on mitigation; recast climate change in terms of clean
energy or "green collar jobs;" and redefine the debate as a matter
of public health. Each of these new meanings, when additionally
connected to an urban or local community focus, is likely to activate
increased public attention and concern.

Bruna De Marchi
We have to use heroes; we need to use health messages;
and we have to use something that tackles unemploy-
ment or something else that brings you close to what the
people need in their country, in their communities.

Massimo Alesii
At the end of our debate we should invert the notion of
places where disciplines must have a role. Then we can
see that communication is a tool, and communicators
should think about the ethical dimension related to producing communication processes which have engendered
deep distortions and biases in the collective perception of
the climate change problem.

**We have to
use something
that brings you
close to what
the people
need in their
country, in their
communities**

1. *A focus on adaptation.* Many scientists, policy specialists, and advocates agree on the urgent societal challenge of climate change but
emphasize more of an adaptation approach rather than a narrow
focus on mitigating greenhouse gas emissions. These strategies focus on coastal development, hurricane and natural disaster preparation, and building and community design policies that ease energy
consumption. Policy experts such as the University of Colorado's
Roger Pielke have long lamented the absence of this type of focus
from major policy discourse. Yet as Pielke and colleagues wrote in a
2007 commentary in *Nature*, thanks in large part to the focus of the
IPCC reports, the year marked a lifting of the two decade taboo on
serious discussion of adaptation strategy.[6]

2. *The clean energy promise.* A second complementary frame is promoted by Ted Nordhaus and Michael Schellenberger, who stirred
debate among fellow environmentalists with their 2007 book advocating a move away from what they call the "pollution paradigm."[7]
Their communication strategy is to turn the mental box of the
economy in favor of action on climate change. As they write in an
article in *The New Republic*, only by refocusing messages and building diverse coalitions in support of "innovative energy technology," "green collar jobs," and "sustainable economic prosperity" can
meaningful action on climate change be achieved. Environmentalists can rail against consumption and counsel sacrifice all they want,
but neither poor countries like China nor rich countries like the
United States are going to dramatically reduce their emissions if
doing so slows economic growth. For that to happen, we'll need a
new paradigm centered on technological innovation and economic
opportunity, not on nature preservation and ecological limits.[8]

3. *A public health problem.* A final emerging frame is the focus on

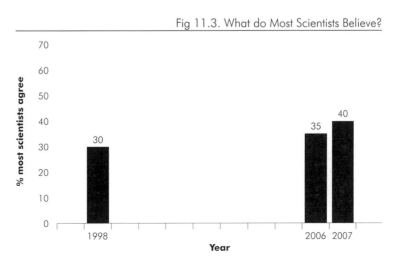

Fig 11.3. What do Most Scientists Believe?

Source: Nisbet & Myers, *Public Opinion Quarterly*, 2007

the public health implications of climate change, emphasizing for the public the connections to infectious diseases, asthma, allergies, or the specific vulnerability of children and older adults. Not only does a focus on linkages to already salient problems activate concern among new audiences, a public health frame also potentially puts climate change higher on various institutional agendas, including many policy contexts where the issue previously had not been given serious consideration. Other than the CDC, examples include the National Institutes of Health, the Surgeon General, new committees in Congress, and/or state and municipal health agencies. Finally, refocusing climate change around public health is likely to bring new organizations and interest groups to the table including medical associations, patient advocacy organizations, or social justice and minority rights groups.

Urban Areas as the New Communication Context
Each of these new frames serves as a potentially powerful lens, bringing alternative dimensions of climate change into clearer focus for segments of the public who may not otherwise consider the issue important. However, these communication strategies also map onto urban and local communities as the new geographic focus for climate change. In the process, for several reasons, communication influence is likely to be amplified.

First, prominent state and municipal officials such as California Governor Arnold Schwarzenegger and New York City Mayor Michael Bloomberg have already taken a lead with policy proposals to address climate change. When these actions are defined as relevant to economic growth or public health, it can be a very important source of opinion leadership for lay citizens while generating additional news media attention to the problem. Moreover, as has been the case with auto emissions or indoor smoking, policy innovation

---

Bruna De Marchi
Raising awareness means behaving in a certain way, not just speaking in a certain way. We tend to interpret communication as words, but communication is mostly action. Recently the European Union funded a very big project on flooding. The second meeting about the project took place in Dresden, in a building that had already been flooded twice. This setting provided a very powerful communication. It is significant that precisely the same politicians or public authorities that on one hand say climate change is an important issue may allow or even favor certain urban developments that are totally contradictory. These are very powerful communication acts without words.

We tend to interpret communication as words, but communication is mostly action

at the state and municipal level can spread across other states and at the national level.

Second, cities are where individuals may be the most vulnerable to the health implications of climate change. Focusing communication around this dimension can help raise the media profile of the issue while mobilizing a diversity of new social groups into action. The vulnerability and differential impacts specific to urban areas may add a new "public face" to the problem, shifting the visualization of the issue away from remote artic regions, peoples, and animals to more socially proximate neighbors and places. In the process, not only do the symbols of global warming change, but the issue begins to cut across media zones, triggering coverage at local news television news outlets and specialized urban media. In all of these cases, the public health focus likely activates concern from audiences who may otherwise rate climate change as a lower tier problem.

Third, urban areas are perhaps the most important audience for climate change communication, since their population density translates into a loud voice on policy and an equally important impact on personal behaviors relative to surface transportation, energy use, and various forms of adaptation.

Finally, because of their population density and culture, urban areas are also uniquely suited for transcending media campaigns and targeting citizens via their interpersonal and digital networks. For several decades, communication researchers have recognized the importance of "influentials" in shaping public preferences, informing fellow citizens, and catalyzing behavior. Influentials are a select number of individuals across social groups who serve as important information brokers. From movies to presidential politics, a small group of citizens typically pay close attention to news and adver-

---

Bettina Menne
Cities can be at the forefront of reducing energy consumption in integrated transport planning, in addressing multiple co-benefits and I will make the case of obesity prevention, heat adaptation, and increasing the equities in the cities.

tising on a specific topic, discuss the issue with a diversity of others, and appear to be more persuasive in convincing individuals to adopt an opinion or course of action. On climate change, if these individuals across cities are recruited and trained to pass on carefully framed information to their peers, the impact of any media campaign is likely to be amplified. In fact, the use of influentials and their social networks is a central strategy in the advertising and media campaign recently announced by Al Gore and the Alliance for Climate Protection.

1. Pew Project for Excellence in Journalism, *State of the News Media*, 2008. Available at http://www.stateofthenewsmedia.org/2008/narrative_overview _contentanalysis.php?cat=2&media=1.
2. R. Cohen, and P. Bell, National Journal Insiders Poll. *National Journal,* February 3, 2007. Available at http://www.nationaljournal.com/njmagazine/nj_20070203_2.php.
3. R.E. Dunlap, "Climate-Change Views: Republican-Democratic Gaps Expand," Gallup News Service, May 29, 2008. Available at http://www.gallup.com/poll/107569/Climate Change-Views-RepublicanDemocratic-Gaps-Expand.aspx.
4. Pew Research Center for the People and the Press, "A Deeper Partisan Divide on Global Warming," 2008. Available at http://peoplepress.org/report/417/a-deeper-partisan-divide-over-global-warming.
5. D. Kennedy, "Breakthroughs of the Year 2007," *Science 318* (2007) 5858, 1833.
6. R.A. Pielke, Jr., G. Prins, S. Rayner, and D. Sarewitz, "Lifting the taboo on adaptation," *Nature* 445 (2007) 597-598.
7. T. Nordhaus and M. Shellenberger, *Break Through: From the Death of Environmentalism to the Politics of Possibility* (New York: Houghton Mifflin, 2007)
8. T. Nordhaus and M. Shellenberger, "Second life: A manifesto for a new environmentalism," *New Republic* (September 24, 2007) 31-33.

CHAPTER 12:   **Governance and Consensus Building**

Marianella Sclavi

The institutional forms of representative democracy and their techno-bureaucratic administrations are ill-suited to the new problems we face in the twenty-first century, inclusive, of course, of the climate change dilemma. Given the global and interconnected world that we live in, devising and implementing public policies that promote a productive economy and healthy society requires forms of participatory governance. Participatory governance facilitates the active political involvement of the citizenry and forges political consensus through dialogue. It improves the capacity of discussion and constructive conflict management. Because climate change imperatives transcend established political boundaries, conflict resolution between new political entities will be more and more essential to policy -making. Fortunately a large body of useful theory and practice of the "Consensus Building Approach" (CBA) has developed over the past two decades, and will be crucial to the new conditions for participatory governance.

I think it is useful to distinguish between two approaches to participatory governance; namely the conventional one, and what is called "Empowered Participatory Governance" (EPG). The latter is grounded in the theory and practice of consensus building in negotiation and mediation. It attempts to advance three general principles in the social sciences and democratic theory. They can be delineated as: focus on specific, tangible problems; involve of ordinary people who are affected by these problems as well as the officials that represent them; develop solutions.

Harriett Bulkeley
I think too often we look for this process of climate governance to be one of consensus. I think the success of the process involves leadership first, rather than consensus. There is almost always a charismatic individual involved. So, probably we need to create lots of charismatic individuals and send them out.

Is it not inevitable that in seeking consensus you always end up with the lowest common denominator? So you may be able to create a plan which everybody buys into, but which has all the difficult bits left out?

The success of the process involves leadership first, rather than consensus

Julie Sze
I think there is a big range within consensus making, rais-

The Consensus Building Approach (CBA) is a method designed to bring people face to face such that they can educate each other about their real interests and search out mutual gains. Empowered Participatory Governance (EPG) grew out of Alternative Dispute Resolution (ADR), which previously was applied mainly to private disputes and negotiations, as an extension of public dispute resolution.[1] In the late 1980s, it gained legitimacy mainly in English-speaking countries, and has since been applied in a number of fields. Environmental disputes provided one of the first arenas in which individual practitioners began exploring the applicability of the CBA to dispute resolution techniques and processes.[2]

The theory and practice of the CBA is part of the larger field of communicative planning theory, which addresses the nature of collaboration, and more specifically defines dialogue as being different from an ordinary discussion or debate.[3] Not all consensus seeking activities are equal and not all should be labeled as consensus building. A number of conditions need to be met for a process to be labeled as consensus building, as follows:

- Inclusion of a full range of stakeholders.
- A task that is meaningful to the participants and carries with it a promise of having a timely impact.
- Participants set their own ground rules for a variety of factors, including individual behavior, agenda setting, making decisions, etc.
- A process that begins with mutual understanding of interests and avoids positional bargaining.
- A dialogue where everybody is heard, respected, and equally able to participate.
- A self-organizing process unconstrained with respect to people, time, or content and that permits the status quo and all assumptions to be questioned.

---

ing the question of consensus versus enlightened authoritarianism. Consensus depends on the situation. Consensus depends on the context where the conversations are taking place, and it depends on the issues. In New York City, under Mayor Giuliani, the municipality's view of communities was that communities were obstructionist. They were against everything, and that was the basis by which the politicians managed communities. In California the scenario was very different. I think that if you have the perspective that communities are obstructionist, you are never going to move ahead. If you study the perspective that the environmental justice advisory group has made on the AB32 (the Global Warming Solution Act), there are very good observations. They highlight serious concerns about waste-to-energy. They wonder why mobile

Consensus depends on the context where the conversations are taking place, and it depends on the issues

• Information accessible and fully shared among all participants.

• An understanding that "consensus" is only reached when all interests have been explored and every effort has been made to satisfy these concerns.

Consensus building aims to produce the following: joint learning; intellectual, social, and political capital; innovative problem-solving; sharp understanding of issues and other players; skills in dialogue; capacity to work together; and shared heuristics for action. In addition, second order effects such as spin-off partnerships and societal and institutional capacity may result. Typically a skilled and trained facilitator is needed to satisfy these conditions. A facilitator conducts a conflict assessment identifying the stakeholders and their interests, to find out which is their "Best Alternative to No Agreement" (BATNA). A facilitator also makes sure that everyone at the table has interests that are reciprocal with some of the other interests. Discovering this reciprocity is the central part of the dialogue. Reciprocity is equally important for the more powerful as well as less powerful stakeholders.

The participatory approaches most commonly used in continental Europe tend toward traditional leanings. "Participation" is seen as allowing more people to enter the public arena, as opposed to rethinking the entire decisional process and the arena itself. Both the English-speaking countries and Europe heavily rely upon the standard debating approach, but in the Anglo-Saxon world, approaches based on consensus building and active listening are more often employed, while in continental Europe the typical and standard debating approach is more often the rule.

The difference between consensus building and ordinary discussion (be it in everyday life or in parliamentary procedures) entails

---

sources are not being considered under the mandatory greenhouse gas report.

Bettina Menne
For climate change– related negotiations, if we could have a legal process based on the principle of majority, like we do have for the international health regulations, we would have had a legal act in place for at least twelve years. But as the whole of climate change negotiations are based on consensus, of course we require a huge negotiation ability and capability. So, you may call it endless negotiations, yes, but at the end of the day, you will have a consensus statement, a ratified implemented convention protocol to follow. We need to keep this in mind.

As the whole of climate change negotiations are based on consensus, of course we require a huge negotiation ability and capability

the passage from a debate-centered decision-making process to a dialogue-centered one. While the purpose of debate is to win an argument, dialogue is about exploring and creating common ground. Consensus building is a dialogue, with a facilitator's help, among people who know that they do not share sufficient goals or worldviews for a conventional type of argument to be effective. Dialogue may be differentiated from discussion or other forms of talk in three distinctive ways:

1. Equality and the absence of coercive influences. Dialogues become possible only after the higher ranking people have, for the occasion, removed their badges of authority and participate as true equals.

2. Active listening. The ability and motivation to respond empathically to opinions with which they disagree or that they find uncongenial.

3. Bringing assumptions into the open. Within the safe confines of dialogue, people can respond without having others pass judgment and challenges.

It can be useful to contrast the search for better communication and decision-making processes in continental Europe and in the Anglo-Saxon world by comparing the communication theory of Jürgen Habermas, representative of the former approach, with the Consensus Building Approach representative of the latter: In consensus building, the discussion does not proceed through the force of better debate, as Habermas suggests. It is not based on reciprocal positions, debating skills, and abilities of reciprocal persuasion.[4] Among advocates of the Consensus Building Approach, theorists and practitioners speak openly of the "missing skills" which are Active Listening and Creative Conflict Transformation. These are necessary to create and sustain the conditions of a dialogical approach to decision making. By contrast, much of the recent talk

---

Antonio Cianciullo

Consensus is reasonable when analyzing this in the global context, when discussing the Kyoto protocol, because that is where consensus is best applied. Consensus is a mechanism envisaged by international law. In cities, however, we have to bring together bits and pieces. So in dealing with climate change at the urban scale, we must build on these parts, these sections of a pathway, and we have to measure consensus on just those sections.

Consensus is a mechanism envisaged by international law. In cities, however, we have to bring together bits and pieces

Claudia Bettiol

In Italian consensus is ambiguous: I rather call it negotiation. Consensus cannot be found in the short term, so if we are looking for negotiation in the short term, we cannot reach a compromise. In the immediate term

Jürgen Habermas	Consensus Building Approach
Looking for ideal type conditions	A pragmatic approach grounded on ADR
Main actors: Philosophers	Main actors: Practitioners
Debate. It proceeds through the force of better argument and is based on reciprocal positions, debate skills, and the ability of reciprocal persuasion. It is the extension of an ideal parliamentary procedure to all citizens.	Dialogue. Participants ask each other questions, learn about the problem and each other, and engage in collective story telling. They tell stories to describe their interests, to imagine what will happen if nothing is done. They search for a future scenario where all their interests are at least better served than they would be if they had not come together.
Questioning taken-for-granted assumptions	Questioning deeply-taken-for granted assumptions

Table 12.1. Comparison between Jürgen Habermas's theory and CBA

and practice of "participatory methods" in continental Europe, and particularly in Italy, is based on the assumption that no particular skill is required to conduct dialogue; that "dialogue is just another form of conversation," and that it does not require a special discipline, but solely content knowledge. In addition, knowing how to "dialogue" is equated with having "debating skills." [5]

Fig 12.1. Participatory Governance – Setting for Public Dialogue & Action

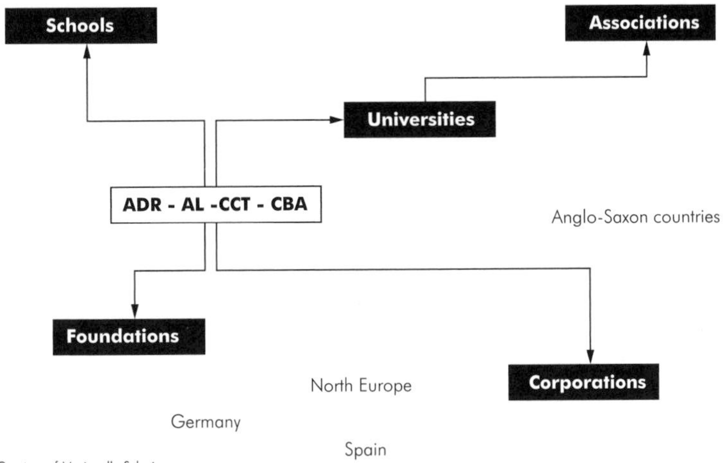

Courtesy of Marianella Sclavi

Most professionals think they already know and possess all of the skills required to get to a good decision and they do not recognize that engaging in dialogue is a highly specialized form of discussion that imposes a rigorous discipline on the participants. It is a discipline as new with respect to conventional procedures as was the parliamentary one, historically, when it developed in opposition to the existing feudal regime at the time. Participatory methods in highly complex and diversified contexts go hand in hand with the learning and practice of dialogic skills.

From a more practical and procedural point of view, the Consensus Building Approach encompasses the following main steps:
- Facilitating group problem setting and solving. This is about generating mutually advantageous proposals and confronting disagreements through the "Active Listening" approach (which goes beyond "a respectful manner"), and entails a joint exploration and enlargement of the range of possibilities. This process draws upon the best available information and ensures that a range of solutions, including some that had not been thought of before, or had been discarded as "impossible."
- Reaching agreement. "Deciding" is not as simple as "voting." It entails coming as close as possible to satisfying the most important interests of the involved parties and documenting how and why agreement was reached.
- Holding people to their commitments. This consideration is more than each person simply actuating what they had promised. It involves maintaining a relationship between the parties so that unexpected problems can be addressed together.

Besides environmental disputes, the pioneers of public dispute resolution have worked in other distinct activity areas including negotiated rulemaking. In the early 1980s, mounting dissatisfac-

---

the only possible form of agreement is a compromise, which is usually of lower quality. Instead, the solution is shifting the advantages over time and pushing them forwards. The idea that we cannot find a solution right now but can think about a future scenario allows for multiple creativity development leading to advantages that might be of financial nature or can be qualitative. Of course, the qualitative aspect, with respect to buildings and the built environment, must entail urban planning study that recognizes those citizens who are going to live there. In any urban regeneration process, what transpires is transformation through private-public partnerships. There is a leader, someone who initiates the transformation process, because he or she sees the economic advantages of this transformation.

In any urban regeneration process, what transpires is transformation through private-public partnerships

tion with federal rulemaking propelled the Administrative Conference of the United States (ACUS) to recommend that agencies try new rulemaking procedures based on the principles of Alternative Dispute Resolution (ADR) applied to large group interaction situations.[6] The Conference envisioned a process by which new rules would be developed through direct negotiation and collaborative fact -finding among all groups likely to be affected by their promulgation. This approach aimed to reduce the time, cost, and acrimony associated with conventional rule making by creating avenues for groups to participate at all steps of regulatory decision making. The decision process held out the promise of producing rules with far greater legitimacy in the eyes of the public, thereby eliminating the endless cycles of litigation that bogged down agency action. This approach to rule -making has acquired more legitimacy now at the beginning of the 21st century, with its results being judged much more effective and stable than those of the conventional approach which was delegated uniquely to lawyers and experts.

Neighborhood and community-based dispute resolution are another area where new approaches to managing differences have been developed and tested. Previously, the practice of mediation was in use, but what is new includes the effort to institutionalize this kind of assistance in centers staffed by volunteer mediators specially trained in the use of emerging conflict resolution techniques. One of the distinctive characteristics of public dispute resolution has been a great capacity for innovation in design of new processes tailored to solving problems, building enduring agreements among people locked in impasse, and improving policy making so that it produced efficient government action responsive to diverse public and private interests. In summary, consensus building is an ad hoc process which focuses on meeting everyone's interests rather than on allowing a majority to rule while the minority remains unsatis-

---

Any kind of project financing initiative is based on the fact that I don't have an immediate advantage in undertaking a specific activity. Instead it is diluted over time, because I draw advantages from the use of that building or whatever facility. So, if we apply these advantages to the urban planning scale, I have to have a negotiation that is diluted over time, and the identification of the time frame is the difficult part.

fied. Mutully advantageous agreements are sought, with the pos-
siblility for a group of organization starting from various divergent
positions and ideas to reach new common solutions highly upheld
by all participants, and greater than their original ideas. Such success is strictly connected with a process which allows every par-
ticipant to put aside traditional means of settling disputes, instead
of embarking on a creative and productive path. It is this realm of
innovation that may prove most useful in terms of building consensus on climate change strategies at the urban scale.

1. R. Fisher, W. Ury, and B. Patton, *Getting to Yes, Negotiating Agreement Without Giving In*
(New York: Penguin Books, 1991); L. Susskind and J. Cruikshank, *Breaking the Impasse:
Consensual Approach to Resolving Public Disputes* (New York: Basic Books, 1987).
2. L. Susskind and A. Weinstein, "Toward a Theory of Environmental Dispute
Resolution," *Environmental Affairs Law Review*, Boston College, 9. 2 (1980): 143-196.
3. W. Isaacs, *Dialogue and the Art of Thinking Together* (New York: Doubleday, 1999); D.
Yankelovich, *Coming to Judgment: Making Democracy Work in a Complex World* (Syracuse:
Syracuse University Press, 1991).
4. J. Habermas, *The Theory of Communicative Action, Volume Two: The Critique of Functionalist
Reason* (Boston: Beacon Press, 1987) and *Moral Consciousness and Communicative Action*
(Cambridge: The MIT Press, 1990).
5. D. Yankelovich, *The Magic of Dialogue* (New York: Touchstone Books, 2001).
6. Americian Bar Association www.abanet.org/adminlaw/news/vol21no2/ams_rip.html
(accessed September 2008).

CHAPTER 13:   **The Mad Max Phase**
Lieven De Cauter

"The behavior mode of the system is that of overshoot and collapse."
Limits to Growth. A Report to the Club of Rome Project on the Predica-
ment of Mankind, 1972

Is it possible that the future of our world looks like some version of
Mad Max, a trashy sci-fi movie in which oil scarcity has turned the
planet into a low-tech, chaotic, neo-medieval society run by gangs?
Is this implosion of the polis, this disintegration of society, not just
probable, but maybe even inevitable? That is the question. In the
famous, *Limits to Growth. A Report to the Club of Rome*, of 1972, this
implosion was called "the collapse of the world system":

If the present growth trends in world population, industrialization, pollution, food production, and resource depletion continue
unchanged, the limits to growth on this planet will be reached
sometime within the next one hundred years. The most probable
result will be a rather sudden and uncontrollable decline in both
population and industrial capacity.[1]

This implosion scenario, so dryly announced in the first *Report to*
the Club of Rome, remains the most probable scenario for the simple
reason that our world is more than ever caught in this logic of
growth - caught, as Sloterdijk has put it, in the hyperkinetic frenzy
of total mobilization.[2] This is to be taken very literally. Of the 16
most air polluted cities in the world, 12 lie in China. One thousand
new cars enter traffic each day in Bejing alone, but only 2% of the
Chinese population owns a car as yet. On the other hand, the ex-

---

Bruna Esposito
I feel that the media is doing a dirty business with catastrophes. In other words, the media seems to really underline only a negative view of things. They put the camera
in the wound only when the blood is running. Once the
patient is healing - you know – and the doctor is having
success, the dying patient is not news any more. So, I
think what is happening now is very dangerous in terms
of the media stressing negativity.

Richard Plunz
The media   The Club of Rome engaged a very Western-centric doc-
seems to really   trine; in the end their only answer was to solve the prob-
underline only   lem by keeping development local; to contain it within
a negative
view of things   the developed world, as a kind of manifest destiny. It is

ample that our group has contributed to the situation by blowing so much kerosene into the air to make this climate conference possible has become a classical but painfully allegorical example by now. The Chinese are quickly giving up their bicycles for cars, just like we are increasingly taking planes. Nobody seems to be able to escape the logic of increasing mobility, of globalization, growth, acceleration. The "essential problem," according to the *Report to the Club of Rome*, was "exponential growth in a finite and complex system." As we are addicted to growth, mobility, acceleration, we are on a collision course with the ecosystem we inhabit.

Since the first *Report to the Club of Rome*, we know emissions have only grown exponentially and keep growing; the demographic explosion is continuing; consumption increasing; and in developing countries it has just started. As a result the ecosystem is more and more destabilized, with the melting of glaciers and perennial polar icecaps as most visible result (besides a sharp decrease in biodiversity, increasing desertification, more hurricanes, floods, heat waves, droughts, etc.). In the two generations since *Limits to Growth*, we have done nothing to alter "'the predicament of mankind" and of the planet. This inertia is not so much the inertia of being too passive, but the inertia of the law of physics: of being caught in an acceleration that will not stop if there is no other body or force to stop it - the inertia of acceleration. In this respect, consider the latest report of the International Panel on Climate Change (IPCC) and the Stern Review, the most recent updates on global warming. When they try to be positive or optimistic, they are not very convincing. Here one of the last messages of the Stern Review: "There is still time to avoid the worst impacts of climate change if strong collective action starts now." But after 36 years of saying this, from *Limits of Growth* onward, nothing or next to nothing has happened, and in fact things have gotten much, much worse. In those thirty-

---

important now to re-examine those issues in world history and to go back at that kind of global vision, but with a more inclusive formula. The truth is that we do not have the institutions to do that. For this conference we decided to limit ourselves to North America and Europe. Limited to this piece of the world, at least we could examine ourselves. From the point of view of the rest of the world there is evidence that, actually, we are not the centre of the universe, both in scale and in intellectual capacity, and many of the issues that we face are being solved elsewhere in ways that, given the constraints of our history and culture, are very hard to comprehend, or even to empathize with.

We are not the centre of the universe, both in scale and in intellectual capacity, and many of the issues that we face are being solved elsewhere

six years we lost for positive action. We have, on the contrary, broken a balance of hundreds of thousands of years.[3]

Besides this "inertia of acceleration," there is another reason why nothing will be done. Naomi Klein has given it a name: the Rise of Disaster Capitalism. The "shock doctrine" was mostly designed by Milton Friedman and the so-called Chicago School. It was first deployed in Chile in 1973, and further developed until it reached it apogee in Iraq. It was at work in New Orleans after Katrina, and in Sri Lanka after the tsunami. Klein sees three elements. The first is a disaster, be it a coup d'état, a natural disaster, or a terrorist attack (like 9/11). Secondly, this shock and its paralyzing effects are used to impose economic shock therapy, toward privatization and deregulation (like the erasing of public housing and public schools in the aftermath of Katrina, or the privatization of security and war after 9/11). And third, there is torture, often with electro shocks, for those who oppose this political and economical shock therapy, from Pinochet to Abu Graib.

We have always thought that relative peace and stability were necessary for commerce and capitalism to bloom. But we were wrong. Naomi Klein has unraveled the puzzle. She has brought together what most of us have always seen as being apart. On the one hand, we have the war on terror as state of exception, and on the other, post-Fordism in the form of the network-society, neoliberal globalization, and market opportunities for growth.[4] The shock doctrine explains certain present-day riddles of the American stock exchanges that were doing so well (at least until recently) while there is war. It also explains why the Tel Aviv stock exchange was not plummeting, but peaking, when Israel attacked Lebanon in 2006. Halliburton, Bechtel, and Blackwater are iconic names for disaster capitalism in Iraq and elsewhere, but the syndrome is much more

---

Bruna Esposito

This past January, in New Orleans, I was invited to participate in a Sunday parade called "Second Line." I knew about Dixieland Jazz in New Orleans, but for the first time in my life I saw a gigantic tuba, walking down the street, in one of the most devastated areas of New Orleans; after Katrina people marching, jumping, dancing along with the band; and there were several stops in front of houses. From each house several people dressed with costumes jumped out, and they were very colorful - everything color and music. But *signori*, what beauty, what energy, what liveliness, what power implied in marching together, singing together, and moving on together to the next stop. I want to say, what a real catastrophe it would be if human beings will not sing any more. Whenever

What beauty, what energy, what liveliness, what power implied in marching together, singing together, and moving on together to the next stop

vast and it is expanding.

The rise of this new form of capitalism is, of course, extremely bad news in the light of climate change and the permanent catastrophe it will entail. It unmasks the argument that capitalism will be forced by the invisible hand of the market (with a little help from politics) to find the tools to mitigate climate change and find ways of adaptation where mitigation is too late. It makes the argument that there is a huge green market out there much less convincing. For this disaster capitalism, mitigation and adaptation are not necessary, and even bad for business, because disaster itself is the business.

The Mad Max phase of globalization will not mean the collapse of capitalism, but rather, the implosion of society. This implosion of the polis, the disintegration of the city (of Man), is not something that is ahead of us, but it is in fact already materializing in the new spatial disorder of the "capsular civilization."[5] Disaster capitalism is making movable green zones in the expanding red zones of the world. Speaking about Katrina, Klein writes:

> At first I thought that the Green Zone phenomenon was unique to the war in Iraq; Now I realize that the Green Zone emerges everywhere where the disaster capitalism complex descends, with the same stark partitions between the included and the excluded, the protected and the dammed.[6]

The new spatial disorder is a collection of plugged-in capsular entities; green zones surrounded by unplugged red zones. These green zones make a layered archipelago, islands inside bigger islands, a sort of Russian doll archipelago, with hotspots where the seams of the seemingly seamless smooth space of the network society are painfully visible. These new iron curtains are the fences and security walls all over the world: in Ceuta, Tijuana, Palestine. To

there will be no one poet left singing a song, then it will
be truly the end of the world.   All the other catastrophes,
I think we can sit down around a table and study them,
find a solution, and work for it.

Marianella Sclavi
We need analytic pessimism and pragmatic optimism.
We absolutely need enthusiasm precisely because the
situation is so disparate. With enthusiasm (which implies
shared values) and a shared sense of a common challenge, we can deal with difficult situations and overcome
huge obstacles. In a crisis we cannot afford to forget that
not just interests, but identities, values, emotions, are all
part of the construction of reality. In this regard, the style
of leadership, is crucial.

keep these worlds apart there is the counter-archipelago of camps: detention centers, labor camps, secret prisons, or outright concentration camp like Guantánamo. All are spaces of exception, some of them not only extraterritorial, but also extra-legal, outside the law - outlaw spaces that keep the split in place, break resistance, and spread fear.

We live in this dualized world already: on the one hand there is the "hyper-reality" of the world of consumption, tourism, media, spectacle; on the other side there is the "'infra-reality" - the unknown, repressed, invisible, ugly reality outside this matrix. Razor wire is the marker of the division line between hyper- and infra-reality. It is likely to be one of the most important features of the architecture and urbanism of the twenty-first century. This is the sort of cyberpunk landscape that will only become more visible and more extreme. And of course ecology is crucial in the shaping of this new dualized spatial order. Air-conditioned, biosphere-like, capsular entities are no doubt ahead of us while the planet heats up. The "green zone" can become a sort of security stronghold as well as an ecological safe haven - the "total air-co" of a capsular civilization.

The Mad Max phase has begun. One concrete recent example encapsulates the perverse logic. To avoid overfishing in European fishing grounds, a grey, often illegal fleet (Chinese, Russian, and other) has completely overfished the West African fishing grounds, so that the fishermen of Ivory Coast and surroundings, having no means of subsistence, take their boats and try to make it to the Canary Islands, i.e. the promised land called Fortress Europe. An estimated 6,000 people died trying to cross in 2007.[7] Ugly as it is, this is just the beginning. Africa, according to the IPCC fourth assessment report, will be among the hardest hit regions by climate change, and the least equipped to adapt. And (not mentioned in

Fig 13.1. The New Spatial Order

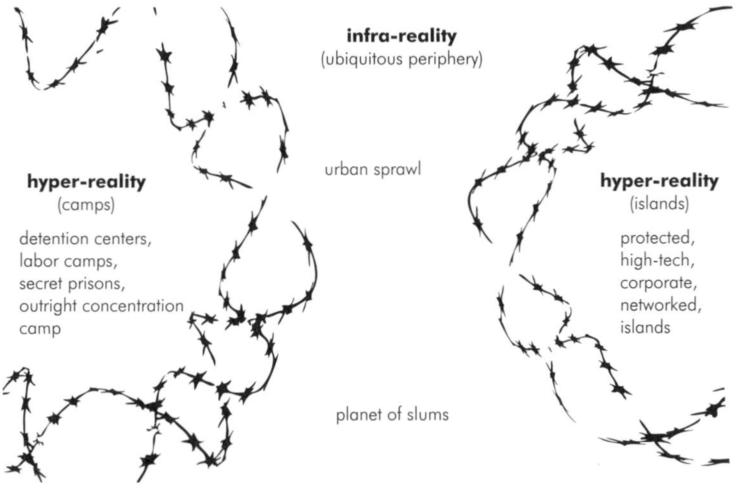

infra-reality
(ubiquitous periphery)

hyper-reality
(camps)

detention centers,
labor camps,
secret prisons,
outright concentration
camp

urban sprawl

hyper-reality
(islands)

protected,
high-tech,
corporate,
networked,
islands

planet of slums

the report) it is also the continent with the sharpest demographic growth. End result: mass migration from Africa to the North, first of all to Europe. This will cause frictions, capsularization, gating, dualization, and eventually the collapse of the welfare system in Europe. The famous Pentagon map of the "integrated core" and the "non-integrating gap" outlines the trace of our walled world. But as the fences are "leaking," the world will become a Russian doll archipelago: smaller islands inside bigger ones; from the scale of continents to the scale of gated communities and high-security condominiums.

If we were to make a graphic map of the New Spatial Order, we could sketch it as follows: There are two global archipelagos that are mirroring each other. On the one hand is the archipelago of protected, high-tech, corporate, networked islands; and on the other the archipelago of camps. In between them is the background; the ubiquitous periphery. This periphery has two faces: the "planet of slums,"[8] the humanitarian disaster of rapid informal growth in the mega-cities in the Global South, and the ongoing ecological disaster of urban sprawl in the North.[9] A neo-medieval, cyberpunk, post-historical science fiction landscape - that is the stage-set for the Mad Max phase of globalization. It is taking shape before our very own eyes.

1. D. Meadows, D. Meadows, J. Randers, and W. Behrens, *Limits to Growth: A Report to the Club of Rome Project on the Predicament of Mankind* (New York: Universe Books, 1972) Summary: http://www.clubofrome.org/docs/limits.rtf (accessed January 2008).

2. P. Sloterdijk, *Eurotaoismus* (Frankfurt: Suhrkamp, 1989).

3. One of the "robust findings" of the 2007 IPCC report is particularly painful in that respect: "Global total annual anthropogenic GHG emissions, weighted by their 100-year GWPs [Global Warming Potential], have grown by 70% between 1970 and 2004." So here we see what happened in these two generations since *Limits to Growth* was published. The time scale is mind blowing, for the report goes on: "As a result of anthropogenic emissions, atmospheric concentrations of N_2O [Nitrous Oxide] now far exceed pre-industrial values spanning many thousands of years, and CH_4 [Methane] and CO_2 [carbon dioxide] now far exceeds the natural range over the last 650,000 years." IPCC, 2007: Climate Change 2007: *Synthesis Report. Contribution of Working Groups I, II and III to the Fourth Assessment Report of the Intergovernmental Panel on Climate Change.* IPCC, Geneva, Switzerland.

4. Klein solved one of the riddles of neo-conservatism, too. On the one hand Leo Strauss and his followers in political theory, and on the other Milton Friedman and his followers in economic theory, or else the political side and the economical side of neo-conservatism.

5. L. De Cauter, *The Capsular Civilization: On the City in the Age of Fear* (Rotterdam: NAI Publishers, 2004).

6. N. Klein, *The Shock Doctrine: The Rise of Disaster Capitalism,* (London, New York: Metropolitan Books, 2007) 414.

7. International Herald Tribune, 2008: "The fish gone, migrants take to sea," *International Herald Tribune*, Jan. 12-13. "World pays a price for love of seafood", *International Herald Tribune*, January 15, 2008.

8. M. Davis, *Planet of Slums* (London: Verso, 2006).

9. This new spatial disorder we have more extensively treated in: L. De Cauter and M. Dehaene, "L'Archipel et le lieux du ban, Tableau de la ville désastre," in *Airs de Paris* (catalogue), Centre Pompidou (Paris : éditions du Centre Pompidou, 2007) 144-148. Also in : L. De Cauter and M. Dehaene, "Meditations on Razor Wire : A Plea for Para-Architecture," in *Visionary Power: Producing the Contemporary City* (cat.), International Architecture Biennale Rotterdam (Rotterdam: NAI Publishers, 2007) 233-247.

CHAPTER 14

The Sociology of Disaster
Bruna De Marchi

Social sciences including geography, psychology, and sociology, have been dealing with risks, disasters, and catastrophes for many decades, each one from its disciplinary perspective and originally with little interest in communicating with one another, let alone with the natural sciences. More recently the trend has been toward closer (although not easy) collaboration between different types of knowledge and expertise, due to the growing recognition that a multiplicity of perspectives needs to be taken into account for the understanding and management of risk issues. Most often the latter are the result of complex interactions between natural and human systems, which can be only partially anticipated. This dilemma exists for many environmental problem -sets; climate change is possibly the paradigmatic one.[1] My reflection is whether and how the insights derived from over 50 years of "sociology of disaster" research with which I am most familiar can contribute to the current debate on how best to address the climate change challenge.

As it can be expected, sociologists focus on exposure of the human system to a certain hazard, rather than on the hazard itself or the physical event/phenomenon at the origin of a disaster. Yet they are fully aware that risks and disasters incorporate natural, technological, and human components, interacting with one another in complex and often unpredictable ways. The traditional distinction between natural and man-made disasters appears somewhat obsolete in addressing human systems vulnerability and resilience to hazards and risks. It is commonly accepted that the two factors are strongly connected, although for both there are multiple defi-

---

In the IPCC report we concluded that millions of people will be affected by climate change in the years to come

Bettina Menne
I said this morning: climate change will cause 150,000 deaths. Now I remember Antonio Cianciullo, at the World Climate Change Conference in Moscow, in 2003, saying to me, "Bettina, what does mean?" I looked at him and said "it means 150,000 deaths." And he said, "No, this means nothing to me. How many deaths will occur actually?" So how much is it actually, this statistic? So we checked. We found it's actually 0.3% of the annual deaths today worldwide. So, it's really not a lot. It's fairly little. And moreover, tobacco smoke today causes more. Transport accidents today cause more. Let's keep this in mind we look at numbers and figures. In the IPCC report we concluded that millions of people will be affected by climate change in the years to come. How? Malnutri-

nitions, and the understanding of their relationship is not undisputed. Notwithstanding considerable differences, all definitions of vulnerability tend to refer to the embedded weaknesses of a certain system, which make it incapable of preparing for, coping with, and recovering from a certain threat or impact. Correspondingly, resilience refers to the ability to do so. In the understanding inspired by ecological studies (namely Holling), resilience appears to be tightly connected with diversity and sustainability, and refers to the capacity of human societies and associated ecological systems to cope with, adapt to, and shape change without losing options for future development.[2]

In this perspective, the key words are diversity; flexibility; organization; learning; interaction; integration of knowledges (in the plural); memory; legacy; innovation; openness to surprise; and adaptation. Whereas some of these terms (e.g. memory and legacy vs. innovation) seem to point to incompatible strategies, indeed all together they highlight the necessity for valuing different perspectives and for promoting a frame of mind that uses lessons from the past to adapt to present contingencies and to increase capacity for foresight. Within this framework, taking into account different types of knowledge is essential, as understanding and insight can originate from all spheres of life, not only scientific work.

Thus adaptation is not a synonym for passive acceptance or resignation. To the contrary, it indicates the creative search for effective strategies for survival, for continued and possibly improved existence, while facing changing (and possibly threatening) circumstances. Understood as a dynamic concept, adaptation encompasses both mitigation and adaptation as commonly construed in climate change jargon, and reflected in the scientific-political arena, where conflicting interests are supported by stakeholders endorsing

---

tion in Africa and Asia could be increasing; many more people suffering from heat waves, flooding, windstorm, droughts. More people suffering from the frequency of cardiovascular diseases, because of changes of our pollution levels in cities.

Why don't we take action? Why don't we react to these issues even though we know they will be happening? One of the risks that we do face today is heat waves. This risk in the Mediterranean will be higher than in the Northern Europe. It's not because we are already exposed to heat, and we will be more exposed. It's simply that we are underestimating the effect of heat on the human body. This issue is quite crucial. The PESETA project of the EU Joint Research Commission has estimated that if the warming

Why don't we take action? Why don't we react to these issues even though we know they will be happening?

divergent lines of research and action. I maintain that the polarization between the advocates of adaptation versus mitigation, narrowly conceived, is detrimental to adaptation in the broad sense,
i.e. as "resilience in action." Strategies addressed only to contain
the "stressor" are likely to be ineffective, especially in the short
term. Although evidence of "man-made climate change" is growing and the fourth International Panel on Climate Change (IPCC)
report confirms that most of the observed warming over the last
50 years "is likely to be" a result of the increasing greenhouse gas
concentrations, many uncertainties still remain on the dynamics of
the overall system, whose physical and human components and reciprocal interactions are largely unknown, unforeseen, unforesee-
able, and uncontrollable, including the long-term effects of policies
adopted to redirect current trends.[3]

For many of those concerned with climate change and its consequences, especially on the environmentalist side, "uncertainty" is a
word to be avoided, as it can serve the cause of those who want
to delay action. Although I am aware of this possible instrumental
use of the word, I maintain that pretending that full certainty exists
is equally dangerous, besides being scientifically incorrect and ethi-
cally questionable. Instead, the key issue to be addressed regards the
bases of policy legitimization, which the modern paradigm, now
somewhat outdated, assumes to be rooted in scientific truth.[4] It is
perhaps time to recognize that we might never be able to base our
choices on "science speaking truth to power." The often advo-
cated "full scientific certainty" may be delayed for years, decades,
and centuries to come, and possibly never be part of our human
experience. Justifying the application of a precautionary approach
on the basis of "lack of full scientific certainty," as in Principle 15
of United Nations' Agenda 21, indicates a lack of awareness of the
existence of unquantifiable and non-eliminable uncertainties.[5]

Fig 14.1. Resilience as a Measure of Sustainability

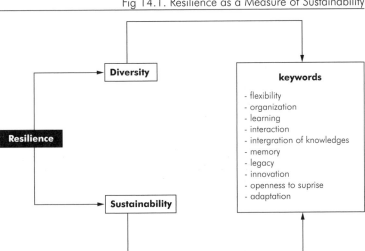

Moreover, improved scientific knowledge may generate more, rather than less, uncertainty as new key variables may be identified or previously unimagined interactions among system components may be discovered. Again, climate change is a paradigmatic example of this scientific state of affairs. As reflected in the very composition of the IPCC, it is now accepted that the issue needs to be addressed from a number of different perspectives, requiring types of knowledge, skill, and expertise not envisaged in the early days, when it was framed essentially as a problem of atmospheric chemistry. Thus, it is not a matter of denying, hiding, or downplaying uncertainty, even if for good cause and with the best possible intentions. It is rather a matter of choosing, through democratic decision processes, how to face uncertainty and ignorance in addressing and managing risk and environmental issues that will affect the health and well-being of our own and future generations.

Creative adaptation, as described in this paper, does include mitigation measures such as actual reduction of dangerous emissions, which can be pursued even in the absence of scientific certainty, based on considerations other than scientific ones. Among these other considerations, attention needs to be addressed to environmental justice, distributional issues, sustainability of present lifestyles, and exploitation of natural resources, etc. Creative adaptation also encompasses measures to contain the negative consequences of events and phenomena that for whatever reason we are unable to control or fully predict. Even in this case our decisions cannot but be made in situations of great uncertainty, which will possibly be increased by those very decisions, whose long-term consequences we cannot fully anticipate. Once again, capacity for foresight and consideration of different possible scenarios are essential, as the relations between the components of complex, tightly coupled sys-

---

continues to increase, we could see 86,000 net extra deaths in European Union countries by the year 2070.* Who is dying? It's people on the fourth and sixth and last floor in buildings. There are much higher rates in several cities around Europe that we've observed. It's those over 65 years old, because of internal physiological reactions to the heat, and the inability of the heart to face the increasing stress. It's people living alone. It's the social isolation of people who are living alone. The perception of thirst over a certain age declines, so if there is nobody who tells you to drink, you just don't drink enough.

* The PESETA Project can be found at: http://peseta.jrc.ec.europa.eu/

If the warming continues to increase, we could see 86,000 net extra deaths in European Union countries by the year 2070

tems are neither linear nor unidirectional and surprise is the rule, not the exception.[6]

Both negative and beneficial effects of change need to be evaluated, and these are likely to be unequally distributed, thus introducing an ethical component in all policy choices. Also, present and future courses of action are largely constrained by past action and learning. When this occurs, it is often post hoc rather than the result of anticipatory design. The 2005 Katrina catastrophe is a recent case in point, which has highlighted our incapacity to learn, to act, and to react effectively, even in a case of a well-known risk and a highly anticipated and largely announced "surprise."[7] There were many others before Katrina, but the peculiarity of this event was the TV coverage of 24 hours a day for many weeks, diffusing *urbi et orbi* the images of "home-grown third world" within today's richest and most powerful country.[8] Hurricane risk in the area was well known (also through repeated previous experience). The Katrina path was continuously tracked and closely monitored. The time and location of landfall were accurately predicted and announced. Never-the-less, an unprecedented "failure of initiative" characterized both pre- and post-impact.[9] This is not the place to discuss the case at length, but suffice it to say that past choices, namely reliance on a levee system, created a "path dependence" that constrains any possible future option.[10] In this "unnatural metropolis"[11], exposed to nearly annual river floods and periodical storm surges, disasters seem to be prompted "by design."[12]

Whether or not climate change will increase the number and severity of hurricanes is not the main point of discussion here.[13] What I consider key, including consideration of the social science studies on disasters, is whether different "tribes" of scientists, experts, risk assessors, administrators, citizens, and all kinds of different stake-

---

Marianella Sclavi

Theodore Roosevelt once said that there are two types of people: those who learn by experience and those who learn by catastrophe. Actually there is also a third type of person: those who do not learn either from experience or from catastrophe. This third kind is becoming a little bit too common. Kurt Lewin taught us that there is a three–step strategy to change deeply rooted habits: 1. unfreeze, 2. change, 3. re-freeze. Too often our communication forgets the first step. We convey change messages without thinking that we have to unfreeze our habits and mutual relationships. That's why the third kind of person is becoming far too numerous.

holders will be able to communicate, promoting a dialogue where different forms of knowledge, perspectives, and experiences can be shared and compared in view of a common goal. Or rather, will they stick to their own limited views, unconcerned or unconvinced of the gravity of future challenges. And of such future challenges, openness to dialogue and collaboration is perhaps the greatest challenge of all.

1. EEA (European Environment Agency), *Late Lessons from Early Warnings: The Precautionary Principle 1896-2000* (Luxembourg: Office for Official Publications of the European Communities, 2001) http://reports.eea.europa.eu/environmental_issue_report_2001_22/en (accessed July 2008).

2. C.S. Holling, "Resilience and Stability of Ecological Systems", *Annual Review of Ecology and Systematics* 4 (1973) 1-23.

3. IPCC, 2007: Climate Change 2007: *Synthesis Report. Contribution of Working Groups I, II and III to the Fourth Assessment Report of the Intergovernmental Panel on Climate Change.* IPCC, Geneva, Switzerland.

4. S. Funtowicz and R. Strand, "Models of Science and Policy", in T. Traavik & Lim Li Ching eds., *Biosafety First* (Norway: Tapir Academic Press, 2007).

5. UNEP (United Nations Environment Programme), *Rio Declaration on Environment and Development, 1992.* http://www.unep.org/Documents.Multilingual/Default.asp?DocumentID=78&ArticleID=1163 (accessed July 2008).

6. C. Perrow, *Normal Accidents: Living with High Risk Technologies (Princeton:* Princeton University Press, 1999).

7. C. Colten and B. De Marchi, "Hurricane Katrina: The Highly Anticipated Surprise", in C. Treu ed., *Città salute e sicurezza: alcuni riferimenti per il governo delle aree urbane* (Milano: Città Studi, 2008 forthcoming).

8. V. Dominguez, "Seeing and Not Seeing: Complicity in Surprise," 2005, http://understandingkatrina.ssrc.org/Dominguez/ (accessed July 2008).

9. U.S. Congress, *A Failure of Initiative Report by the Select Bipartisan Committee to Investigate the Preparation for and Response to Hurricane Katrina* (Washington: U.S. Government Printing Office, 2006) http://www.gpoaccess.gov/katrinareport/mainreport.pdf (accessed July 2008).

10. C. Colten, *An Unnatural Metropolis: Wresting New Orleans from Nature* (Baton Rouge: LSU Press, 2005)

11. Ibid.

12. D. Mileti, *Disasters by Design: A Reassessment of Natural Hazards in the United States* (Washington, DC: Joseph Henry Press, 1999).

13. A.R. Pielke Jr. and D. Sarewitz, "Bringing Society Back into the Climate Debate," *Population and Environment* 26, no. 3 (2005) 255-268.

CHAPTER 15: The Antarctica Project

Jorge Orta

As an artist, for 35 years now, I have dealt with complex societal problems and in order to do so, I have worked within a mode of "creative complexity" integrating maximum numbers of people. In fact, I am convinced that this world requires complex and interdisciplinary approaches in order to answer complex situations. Artists cannot respond in isolation to the changes taking place in the world. Our response has to be elaborated on a collective basis, and in fact I collaborate with philosophers, sociologists, and scientists in order to shed light on the changing human condition. The Antarctica Project explores the climate change issue, among other things - and the question of the displacement of people that it will cause.

For an artist, asking questions is to ask oneself about the factors influencing collective and individual behaviors. How is it possible, to explain the increasing loneliness, although we are always surrounded by people? How is it possible to explain fear, when "40 Cities" was born to reassure us (C40 Climate Leadership Group)? How is it possible to explain precarious urban conditions when the city concentrates all our means? How is it possible to explain anonymity and indifference when we are all interlinked and everything leads us to think that abundance is impoverishing us; that communication is stunning us; that concentration is isolating us; that the system of security is generating fear and therefore we no longer trust other people, our neighbors or our families? Facing the increasing pressure of people obliged to go in exile, can we the privileged, return to ourselves, behind the frontiers and boundaries? Sometimes boundaries are broken by the lack of hope, by

Massimo Alesii

Maybe we should place this debate within a broader alignment, not just thinking in emotional terms, or in economic terms. We should bring this debate to the ethic bottom line of the common good.

Climate change, for sure, is a mobilizing intercultural and cross-sectoral phenomenon

Climate change, for sure, is a mobilizing intercultural and cross-sectoral phenomenon. It is something that strikes deep within and also on the surface. It interacts with every level of our society and of all societies. In this regard, it leads to a participatory approach, an active participation. For sure, cities and city means communities. Cities could be the catalysts, the places where participation coalescences. This is what cities were created for and if we go back to the political "locus," cities are the places

people who have nothing to lose.

The Antarctica Project is our metaphor for the current world, for the fragility of today's world; and at the same time, for the displacement of people following climate change among other things. We have been working on this for fifteen years, accepting the action/reaction delay over time - that each intervention somehow leads to long-term consequences that prevent us from immediately understanding the cause and effect.[1]

The Antarctic Treaty Project

The Antarctic treaty was signed in 1959 by twelve countries: United States, Argentina, Australia, Belgium, Chile, France, Great Britain, Japan, Norway, New Zealand, South Africa, and Russia. They have transformed the sixth continent into a shared continent for scientific exchanges. For the first time in the history of mankind this treaty has opened the door to a new utopia. The Antarctic is the land that respects environment. It is demilitarized and is open to everybody, based on scientific exchange. Antarctica is the sixth continent, covering an area of 14 million square kilometres, with mountains and glaciers. This area comprises two territories: Western Antarctica with archipelagos linked by ice at the south extremity of the Andes; and Eastern Antarctica with a great plateau covered by ice caps, and the sea with impressive floating icebergs. Here we have a deep, silent, wonderful ice culture with 80% of all the water captured in ice caps.

We decided to set up villages in Antarctica as a symbol of hope. For some years now, we have created sculptures that can be integrated to create a kind of village, which is a metaphor for nomadic architecture in response to precarious living conditions that dominate our societies. But we don't only work with metaphors, we actually

---

where decisions are being made, where decisions are debated, where negotiations unfold and this is conducive to the common good. This is the ethical dimension of the entire process.

Cinzia Abbate
It is frightening to see this image of the sequences of the arch through the tents and thinking that perhaps this could be the new scenario of Venice or, may be New York. This collaboration about emergency dwellings, emergency village is starting to be the new topic of the design also for urban planners.

Giuseppe Tripaldi
As biologists, we are looking at nature. We are looking at

Cities are the places where decisions are being made, where decisions are debated, where negotiations unfold and this is conducive to the common good

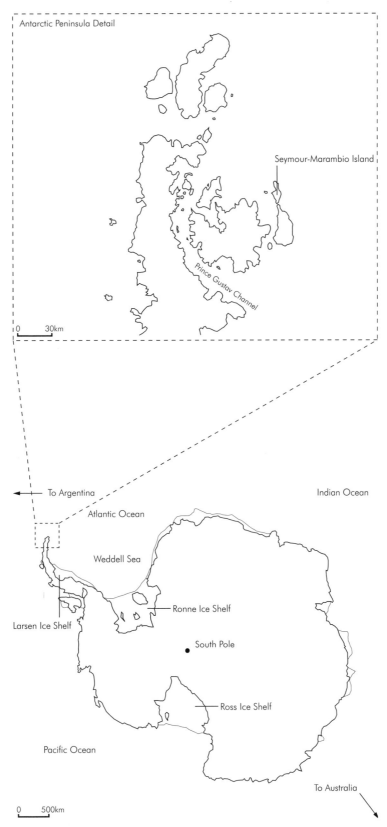

Fig 15.1. Antarctica Project Siting

work in real environments, expanses of land occupied by people and animals - a refuge, an asylum, a place where you can actually go to be safe. Nations are human communities characterized by the conscience of their own historical cultural identity as a language or religious unity. Community is defined as a political entity located on a particular area and institutionally organized as a state.

The Antarctic Village Project

We started with the creation of four villages. The location is at the Marambio Base on the Seymour-Marambio Island (64°14'S; 56°37'W). Installation of the South Village was on Friday, February 23, 2007, the first Antarctic village in this sector. The first day was quite good for this kind of operation. Visibility was between one hundred and nine hundred metres with snow and fog, and the temperature was -9°C, with winds at 12 kilometres per hour. The North Village was installed on Saturday, March 24, 2007 and was the second camp. Twilight was for 3:30 in the afternoon and sunset 9:23. The wind was southwest at 24 Km/hr; temperatures between -4 to -14°C with snow and fog. So you can see - an extreme reality.

The Antarctic Passport Project

Free circulation, free movement is the key point, and so is the terrible tragedy of people displaced. I believe that this is a major threat to fundamental human rights, and we know that free circulation has been recognized by the Universal Declaration of Human Rights in 1948. One of the basic problems of our time is that the status of people is sometimes not as important as the status of money, capital and trade, networks, diffusion of production centers, competition and sports, communication, satellites, tourism and organized crime. And therefore we proposed to amend the Universal Declaration of Human Rights with an additional Article 13.3, that would read:

---

the environment and biosphere. These are very complex systems, and therefore there are not simple solutions for these complex problems. This is why we can't easily come up with shared decisions for such fundamental questions on the future of our society.

Lieven de Cauter
Let us not produce another book. Let's produce a petition which is worldwide, a manifesto, which is free and available and signed by any Nobel prize winner or other authority we can find. It is easier than you think.

Courtesy of Jorge and Lucy Orta

Fig 15.2. Antarctica Project Shelters and Clothing

Every human being has the right to move freely and cross frontiers to their chosen territory. Individuals should not be deemed of an inferior status to that of capital, trade, telecommunication and pollution, all of which have not boundaries.

We created the Antarctic passport. The new citizens of the world need to protect the planet, need to fight every act of barbarism, misery, and terrorism. They need to promote social progress. Human dignity has to be respected and unalienable rights to freedom, justice, and peace have to be defended in this world. We have mobile units to deliver these passports. In order to obtain the passport you have to have your ID taken and to be introduced by seven people.

The Antarctic Flag Project

This idea came on the occasion of the aboriginal celebration of America's Incas between June and October 1992. We were in the Andes and we witnessed the meeting of two hundred thousand Incas who were meeting in Cuzco. We saw that these cultural meetings have many shadows, so I was wondering what we would be doing with Antarctica in twelve hundred years. That is why our Antarctic flag has all the flags of the world. It is like a prism concentrating all the national colors, each national identity is placed one together with the other, hand in hand with the merging of all the edges, as a symbol of a single identity.

This is the way in which we are working. We are using all means and allowing different forms to find a nexus - a relationship between individuals and among individuals, to give strength to the group. And finally, I would like now to say that for decades we have defended catalyzing an art which is able to produce synergies and reactions that can lead to change in society. This is about nurturing Utopia, and the possibility to formulate a dream. We are trying to create a new utopia - a prototype Utopia, so there can be a real ability to implement a new social contract.

---

1. For an extensive description of the Antarctica Project, see:
http://www.studio-orta.com/artwork_list.php

CHAPTER 16

A Vicious Circle
Cinzia Abbate

It seems to be a vicious circle. Cities are growing impressively and helping to break the fragile balance of microclimate and macroclimate. In response, the resultant climate change threatens to devastate the cities. Al Gore's film *An Inconvenient Truth,* reminds us that if the ice sheet of Greenland melts, forty million people would have to leave the Shanghai area because of the consequent flooding. Urbanization is down-zoning agricultural and forest cover, with resultant modification of the geographic and topographic features of the Earth, which in turn must modify our traditional city planning methods.

Traditional urban planning methods used by architects and town planners are no longer safe and efficient. The simple topographic maps used to set the trajectories for urban futures are not reliable enough without interpolation with maps produced by climatologists using scientific models. Indeed, this new global cartography must demonstrate the ongoing geographic transformation processes caused by global warming. It must be a powerful and revolutionary geographic device. To understand this imperative, suffice it to think about how the international real estate insurance companies are already deploying these considerations to assess global urban areas.

Urban planning, therefore, is becoming a complex and multilayered interdisciplinary process that concerns not only specialized professionals but also scientists, climatologists, economists, sociologists, doctors, journalists, philosophers, artists politicians, and, last but not least, concerned communities. Architectural design, together

---

Claudia Bettiol
Maybe, the most important one, in order to trigger the process, is economics because if economists realize the benefit of the process of transformation, the easier it would be to achieve it. In order to disseminate this type of culture we signed an agreement that will be in force next year. The agreement was with Rome University La Sapienza, with the involvement of the others universities in the region: all the new students enrolled in all the schools of these universities will have two mandatory credits related to environmental and energy sustainability.

If economists realize the benefit of the process of transformation, the easier it would be to achieve it

Harriett Bulkeley
In the UK the government campaign has moved away from exhortations to individuals' action, toward empha-

with advancement in building technology, is succeeding in making more and more energy-efficient and independent buildings, often built with eco-friendly materials. By contrast, urban planning faces unresolved challenges in identifying new sustainable urban models. By the term "urban sustainability" not only do we mean the balance between natural and built environment, but also the social and economic balance that future cities will have to manage. This interpretation unveils the importance of devising new progressive urban planning methods and political strategies on all geographical scales:

> As the world urbanizes, cities are being globalized. Not only is urbanization increasingly reaching everywhere…everywhere is increasingly reaching into the city, contributing to a major reconfiguration of the social and spatial structures of urbanism and creating the most economically and culturally heterogeneous cities the world has ever known. [1]

Cities cover less than 2% of Earth's land surface but they account for some 75% of global energy demand and they produce 80% of CO_2 and greenhouse gases. "In 1975 there were only five megacities with more than 10 million inhabitants. In 2007 there were nineteen. There are projected to be 27 in 2025, 22 of which in Third World countries."[2] In China alone, 46 cities have passed a population of one million since 1992, making up a total of 102 cities. In the United States there are only nine.[3] Globally, many cities have expanded along the coastal line and estuaries; as a consequence they have to tackle sea level rise. "Eight of the world's most populated cities are close to seismic areas, some of them are located near volcanoes, and six of them are vulnerable to hurricanes."[4] Europe alone imports nearly 50% of its energy needs and the trend line is expected to increase to 70% by 2030. The uncertainties are enormous. Economic planners and politicians still have no strategic

---

sizing collective behavior. There is one campaign that says it is all about everybody doing 20% - reducing the energy by 20% - so you drive to work four days out of five or you turn your thermostat down one degree, and the idea is that everybody is doing something similar and that it all adds up. Two initiatives - one is called "Manchester Is My Planet," and another called "Carbon Neutral Newcastle" - are a mixture of public and private authorities trying to get across the message that Manchester, collectively, can do something about climate change and so can Newcastle. There are lots of variations: "Oxford is My World," "Stoke is My Green City," and so on and so forth. This approach seemingly is a growing phenomenon in the U.K. Has this collective frame caught on elsewhere?

In the UK the government campaign has moved away from exhortations to individuals' action, toward emphasizing collective behavior

position on sustaining the energy production needed to sustain this new wave of urbanization.

By 2050, with 75% of the world's population living in urbanized environments, we need to ask what shape will these cities have? What will their optimal density or spatial configuration be? What will their streets, public spaces, and buildings look like? What kind of social structure will they be able to maintain? Which new kind of governance models will be able to run these new entities?

Our present prospect for urban adaptation through design is not so promising. The international "star system" as an incubator for urban design innovation has its severe limitations.[5] With the exception of elegant formal solutions, innovative transportation systems, and alternative power generation technologies, some new cities such as Masdar in Abu Dhabi, designed by Foster and Partners, simply follow an old city model, enclosed within walls. As a consequence the growth potential is frozen and the challenge of social inclusion is disregarded. Tri-dimensional concepts of public space refer mainly to commerce as characterized by an urban elite and its consumerist needs. As Saskia Sassen has written: "The currently dominant economic globalization emphasizes hyper mobility, global communications, the neutralization of place and distance."[6]

The Dubai Eco City and Dubai Waterfront, which were designed by Rem Koolhaas to compete with Foster's Abu Dhabi scheme, also take cues from the same model of an enclosed city, prompting one critic to point out that it is "… architecturally stupendous yet profoundly exclusionary. Does its compact size make it easier to seal off from supposed undesirable?"[7] Falk, the new district designed by Renzo Piano and Carlo Rubia for the Milan, Italy region, manages to conceive a "green" city that is technologically advanced

---

I arrived there, and people were "adrenaline," full of life, full of wishes, really wanting to live and to be there

Bruna Esposito
Talking about pre-judging New Orleans, before my arrival I was full of preconceptions and fears because the picture I had was dominated by catastrophic news. Once I arrived, I found another city, a city which is alive with people motivated to stay, people wanting to live there, and those who were able to stay - wanting to continue to stay there and rebuild. And, moreover, before going my feeling was: oh, *poverini*, oh, poor them. And I arrived there, and people were "adrenaline," full of life, full of wishes, really wanting to live and to be there. I often saw a sign "Build or Leave." And what I learned from this experience was incredible, working at the Contemporary Art Centre for an exhibition that was titled, "Something from Nothing." We were asked to make art without buy-

and clean, and yet it was especially targeted to the upper middle class. Notwithstanding such attempts at design innovation, the figures are unequivocal: one person in six lives in decayed suburbs. Where and how will these people live? Keeping in mind that if geography continues changing at its current pace, some global cities are at risk of disappearing or themselves becoming big, flooded, and decayed suburbs.

The 20th century proliferation of town planning practice is unprecedented in history, and the propagation of roles and rules has often hindered development and innovation, thus freezing the cities' potential. The way cities have been conceived, through segregating their functions, homogenizing populations and through zoning, depriving the communities of the chance to realize their potentials, has prevented communities from finding the space and time needed to develop and adapt themselves. Embodied has been the concept of society as a closed system, linked to the constraints of capitalism. Now town planners and architects are having a hard time designing an alternative basis for the staging of urban evolution and transformation. An exaggerated persistence of mainstream city planning practice tends to alienate the city dwellers' bargaining rights, and prevents them from acknowledging real problems and their solutions by means of an increased sense of duty.

Our present era, which is marked by social insecurity and a sense of incompleteness (rather than mere fear), must be regarded as a boost towards a new kind of urban planning, capable of triggering the creation of new kinds of informal economy, able to intertwine differences that can lead to something new and particular. The research on urban competitiveness described earlier by Matteo Caroli highlights the importance of particular urban characteristics and intangible assets, of creating new kinds of economies and new

---

ing anything. So, it was professionally quite challenging, but, on the other hand, it was simply human, the sharing and helping.

Antonio Cianciullo
At the end of the 1800s London could not breathe because of the horse manure. The stench was horrendous and there were risks of terrible epidemics coming from this. But horses did not disappear because people were scared of manure. Instead they disappeared because someone invented the automobile, seizing the opportunity to say that manure stunk, they used that to sell cars. We have to get rid of the chemical manure which is polluting emissions and find a cleaner way.

quality standards.

The current global drive towards urban agglomeration is the first stage of a new level of global economic development. Problems linked to urban poverty and inequality will lead to the creation of 21st century local and global policy beyond past precedents. "One of the main future challenges of town planners primarily consists in finding a way to exploit the economic power of regions and the creative potential of dense and heterogeneous urban populations. Its second purpose is to control the conflicts originated by the ever-increasing economic and social inequality, cultural diversity and global warming. It is therefore crucial to set up the proper local and regional governance."[8]

It is vital for every specific architectural or urban planning endeavor to be catalyzed by the expression of human activity – to be capable of interacting with the dynamism of urban agglomeration and its power to generate creativity, innovation, and economic development. How then can cities convey a sense of order and security to their inhabitants? New urban planning can no longer settle for reassuring aesthetic solutions by "privatizing" the city. The concept of public space goes hand in hand with the right to adequate habitation, a social contract that transcends individuals and unites them by means of a mutual agreement based on shared principles.

Who controls the social and spatial transformation processes of city districts? Can architects and town planners contribute to the creation of public safety in disadvantaged areas? If left on their own, of course not. What we need is a shared planning activity that connects converging points of view. Urban risk reduction must be accompanied by the objective of poverty reduction, which must be associated with international engagement in improving the living

Fig 16.1. Environmental Risk Planning

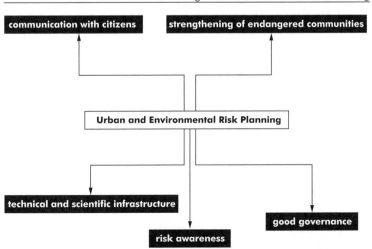

standards of these people. This is just the first step, and then there is still a lot to be done. Good urban and environmental risk planning must be based on more widespread risk awareness, on good technical and scientific infrastructure, on clear and plain communication with citizens, on the strengthening of endangered communities, and on good governance.

1. E. Soja and M. Kanai, "The Urbanization of the World", *The Endless City* (New York: Phaidon Press Inc, 2007), 54.

2. United Nations, *World Urbanization Prospects. The 2007 Revision* (New York: United Nations Population Division, 2007).

3. T.J. Campanella, *The Concrete Dragon. China's Urban Revolution and What It Means for the World* (New York: Princeton Architectural Press, 2008).

4. Z. Chafe, "Reducing Natural Disaster Risk in Cities," *State of the World 2007* (New York, London: W.W.Norton & Company, 2007).

5. Forum for Urban Design, *A Critical Situation: What to Make of Starchiecture.* (New York: The Forum for Urban Design, Inc., 2007).

6. S. Sassen, "Seeing Like a City," *The Endless City* (New York: Phaidon Press Inc., 2007) 277.

7. N. Ourousoff, "City on the Gulf: Koolhaas Lays Out a Grand Urban Experiment in Dubai," *New York Times* March 3, 2008. http://www.nytimes.com/2008/03/03/arts/design/03kool.html

8. E. Soja and M. Kanai, "The Urbanization of the World", 68.

CHAPTER 17

Uncertainties
Alessandro Lanza

When speaking about mitigation and adaptation it is important to bear in mind the elements of scientific evidence that can show us important areas of our scientific ignorance. In my opinion, there is an interesting point that this debate has tended to neglect; that is, the close relationship between greenhouse gas (GHG) emissions, concentrations in the atmosphere, temperature increase, and consequent environmental damage. It is well known what emissions and GHG concentrations are, in the sense that it is a relatively simple issue: gases are released when fossil fuels are burned. Nevertheless, the climate system reacts to GHG concentrations in the atmosphere but not to emissions. Indeed, this triggers the first problem; we switch from emissions to GHG concentrations. Of course there are some models we can use, and we somehow know there is a relationship between emissions and GHG concentrations. Greenhouse gas concentrations, for which there is well-known data, are increasing impressively (figure 1). These data on GHG concentrations are based on snow probes; this is how we know about past emissions.

Things get more complicated when we get to the next step, which goes from GHG concentrations to atmospheric temperature. The IPCC Report[1] has set a 2 to 2.5°C range, depending on the scenarios, as the maximum increase in temperatures that the Earth can bear. Nevertheless, the relationship between greenhouse gas concentrations and temperature is far more complicated than the one between emissions and concentrations. We know that we burn fossil fuels in an incorrect way, thus releasing GHG concentrations. We also know that GHG concentrations in the atmosphere are in-

---

Lieven De Cauter
The proposition of the climate rumor or the certainty of uncertainty - whatever you would call it - is one of the major reasons that people including — especially — businesses or government, will not act. For me, that is like - it's devastating!

Matteo Caroli
Business works on uncertainty. So it might be not so relevant what is the precise of likelihood of warming
Business works on uncertainty. So it might be not so relevant what is the precise of likelihood of warming; whether it is 1.5°C or 1.6°C, is not actionable from the business point of view.

Lieven De Cauter
I think that you misread the IPCC report if you stress un-

creasing, but we don't know much about the nexus between the increase in GHG concentrations and temperature increase.

Climate Change

- Water resources
- Agricultural and food security
- Terrestrial and freshwater ecosystems
- Coastal zones and marine ecosystems
- Human settlements
- Energy and industry
- Insurance and other financial services
- Human health

Stresses

Impacts

Courtesy of Alessandro Lanza

Fig 17.1. Climate Change and Environmental Impacts

A chart summarizing several common models well known in the literature shows how these models can simulate an increase in temperature that is equivalent to specific carbon dioxide ppm levels in the atmosphere. The IPCC set stabilization levels to 550 pm carbon dioxide equivalent - that is, the key figure. Nevertheless, the expected 2°C rise in global temperature at that level of concentration shows an extremely wide oscillation, which highlights our scientific uncertainty on this issue, not that I think this uncertainty should prevent us from taking measures. This is far from what I think. But uncertainties exist and there is a lot to be done. If we want to stick to what we already know, science tells us that the uncertainties on particular phenomena (such as the relationship between GHG concentrations and temperature increase) are still too many. We know

certainty, because the evidence of certainty given in the report is overwhelming. But speaking as a philosopher in the abstract, uncertainty exists everywhere. It exists in mathematics and in logic. In real life there is always uncertainty. Someone walks into the room and says, "There is a 50% chance that this roof will tumble upon our heads; we need to get out the room." And then somebody says, "Are you sure?" It is not sure, but it is likely. For our purposes I think the word "uncertainty" should be banned. We are so certain that we can even discuss, academically, the amount of human impact on climate - the certainty. Even in a sort of safe haven like Belgium, record temperatures have been reached in 2006 and again in 2007. It was the record of all times. So, I think, we owe it to ourselves, as a sort of scientific certainty, to ban the

uncertainty exists everywhere. It exists in mathematics and in logic. In real life there is always uncertainty

Fig 17.2. Global Atmospheric Concentrations of Three Well-Mixed GHGs

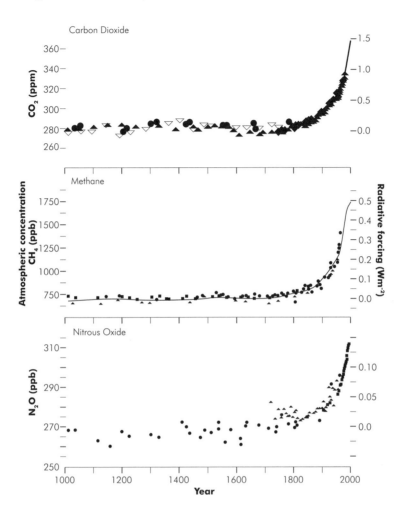

Source: IPCC Fourth Assesment Report, Working Group I Report The Physical Science Basis, 2007

Fig 17.3. Global CO$_2$ Emissions by Sector

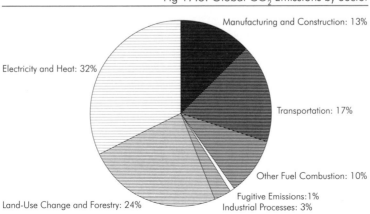

Manufacturing and Construction: 13%

Electricity and Heat: 32%

Transportation: 17%

Other Fuel Combustion: 10%

Land-Use Change and Forestry: 24%

Fugitive Emissions: 1%
Industrial Processes: 3%

Note: Transportation emissions include international transport emissions, referred to as "international bunkers"
Source: World Research Institute, Climate Analysis Indicator Tool, 2006

very little about the relationship between temperature increase and the consequent environmental damage; to be more precise, science is lagging behind the phenomenon. When speaking about the main greenhouse gases, such as carbon dioxide, and also other gases, the impact of one kilogram of carbon dioxide is not equivalent to one kilogram of methane. These substances have a different global warming potential, therefore these gases have to be regarded as a whole.

We often have a distorted view of polluters. Common wisdom always points the finger at the industrial sphere, but in reality it is the energy sector – the production of electricity – that produces the greatest amount of carbon dioxide among all industrial sectors. If we have a look at the International Energy Agency reports, we understand that over the next 30 years China and India will produce energy from coal, thus increasing the emission share attributable to the Energy Sector. Notwithstanding all this, non-energy-related emissions are noteworthy: land use accounts for 18%, agriculture for 14%, and waste accounts for 3%.[2] If we take a close look at the OECD contribution, we will notice that Europe's demand for energy has remained relatively constant compared to that of developing countries, which keeps increasing exponentially.[3]

We know almost everything about energy-related data, since it is about a chemical and stechiometric relationship, and consequently it is relatively simple to calculate carbon dioxide emissions from energy demand. On the other hand, there are still too many uncertainties on data regarding non-energy-related emissions, and that is why we are currently trying to broaden our competence in this area of study.

The environmental impacts of climate change are complex mech-

---

word "uncertainty." Humanity should take immediate action, which means we should be alarmist. If we are not alarmist – and that is out there since 36 years from the Club of Rome. So, for 36 years we have not been alarmist. So, if humanity does not shift to emergency gear, we are really not only damaging humanity or human race but, even more so, for hundreds if not for thousands of years, the planet, bio-diversity, etc. So, this is a plea and will be a continuous plea for alarmism and for rehabilitating alarmism. I think, for me it is very clear: if we don't get very concrete actions that alert people, changes people's habits very concretely, like now, we can just as well lay back and do nothing.

anisms that we know very little about: from erosion and coastal flooding to potential agriculture changes; from consequences for public health to water resources; from ecological changes in our forests to changes in the industrial and energy sectors. Climate change impacts are both "physical" and "economic." We often hear about the economic consequence of physical impacts, such as the negative effects coastal erosion has on tourism, or agricultural production hindered by desertification. The scientific community is working intensely on these much-debated issues, but this does not yet result in clear data.

The Stern Review outlines the environmental impacts of climate change. It explains what the environmental impacts are, and it assesses what economic costs they would entail if immediate measures were to be taken, and the costs of not taking such actions. Environmental impacts are a complex matter, since they include both direct and indirect effects of human action on natural and nonnatural systems. The Stern Review turned out to be very useful; and indeed, I am using it as a matter of discussion in this paper. Nevertheless, the Stern Review drew a lot of criticism from several points of view, for example, around the question of discount rate. Discount rate is commonly used to evaluate future flows of costs and benefits and is an important index. The values taken into consideration change completely if the discount rate is 3% instead of 6%. The Stern Review used a 0% discount rate; in other words this means that a thing will preserve its present value in 100 years. This assumption is hard to accept from an economist's point of view. It means that you are prepared to accept 100 dollars today, and the same amount of money in six years' time - that this choice makes no difference for you. In economics, you shudder at the bare thought of transferring money to the future without a discount rate.

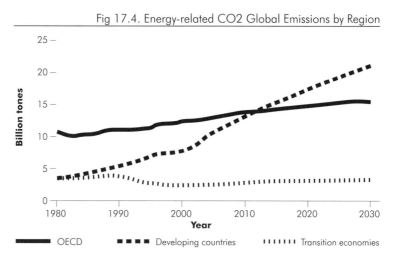

Fig 17.4. Energy-related CO2 Global Emissions by Region

Source: International Energy Agency World Energy Outlook 2006

In conclusion, global warming is still enveloped by great uncertainty. We have been learning much about global warming in the last twenty years, but the issue is still highly controversial. From a geographical point of view it is a complex matter and developing global warming policy in itself is a complex problem in that the damage suffered by a country is not necessarily directly linked to the emissions it produces. That is why international consensus and cooperation is essential. If every country were directly choked by its own GHG emissions, this would lead to a domestic political drive towards proactive policies. But since it is a global problem and countries do not necessarily directly suffer the consequences of their own emissions, it is a hard issue to tackle. With reference to the Italian situation, the country will produce about 3% of world emissions in 2030. Should Italy decide not to produce emissions, or rather to produce 3 or 6% of emissions, this would not change the situation from a climatic point of view. As a consequence, Italy now has to take measures without weighing upon global emissions, even if the economic cost for an emission reduction were extremely high and the consequences for the country severe. This last issue is a time-projection problem: while climate change policy needs a long-term approach (30, 40, 100 years), politics, just as recent history has taught us, too often focuses on short-term issues.

1. IPCC, 2007: Climate Change 2007: *The Physical Science Basis. Contribution of Working Group I to the Fourth Assessment Report of the Intergovernmental Panel on Climate Change,* Cambridge University Press, Cambridge, United Kingdom and New York, NY, USA.
2. S. Putt del Pino, R. Levinson, and J. Larsen, *Hot Climate, Cool Commerce: A Service Sector Guide to Greenhouse Gas Management* (Washington: World Research Institute, 2006)
3. The International Energy Agency (IEA): World Energy Outlook (2006) 70, 11.

CHAPTER 18

Out of Alternative Explanations
Antonio Navarra

In recent years the picture of the growing global planetary surface temperature has become familiar to many of us. It is basically a symbol of the warming planet in the same way that the Wall Street index indicates changing moods in the global markets. It is well evident that the surface temperature made a marked turn in the last century with a drastic rise in the last 20 years. The data is expressed as deviations from the long-term 1960–1990 average. The period after 1980 has seen the same increase in temperature (about 0.4°C) as the period 1900–1940; record after record has been established after the '90s, and the decade 1990–2000 shows officially as the warmest on record.

Volumes have been written discussing this picture, but the main problem is to assess the possibility that this remarkable behavior may be just a casual fluctuation of the climate system. Climate entails a turbulent character that has the habit of changing widely. What are the chances that a fluctuation in climate may actually fluctuate too much? There are ways of quantifying the probability that a certain deviations from some reference level, like the 1960–1990 average, may indeed come from the turbulent nature of the system.[1] This is very common in science any time you cannot find a simple, direct cause and effect, sometimes because you have not identified all the important factors in the problems or there are simply too many of them to investigate one by one. It is customary to accept such results if the probability of a casual occurrence is 1 to 5%. For the global climate case, the casual explanation is that the increase in the temperature is due to "natural" variability, whereas

---

Antonio Navarra

I find the discussions about the scientific credibility of climate change a little quaint. We had a very strong scientific debate. It went on for several years. What we are now listening to in the media is not a scientific discussion. The fact that one is using a scientific vocabulary does **Most of what** not make a discussion scientific. The fact that a paper **we are labeling** has appeared does not mean that it is true or correct. **as a scientific** Important is where it is published, who published it, which **debate on** method was used. Most of what we are labeling as a **climate is** scientific debate on climate is actually manipulation of **actually** symbols and concepts and manipulation of human psy- **manipulation** chology. So, it is not science, and the discussion is often **of symbols and** useless and quickly transforming into an afternoon talk **concepts and** show. **manipulation**
of human
psychology

the external or causal explanation is that there is a changing factor in the composition of the Earth's atmosphere that is altering the surface temperature–regulating mechanisms. The greenhouse gases, carbon dioxide, methane, and the others are the obvious suspects, as they are principally responsible for maintaining the surface temperature and because their concentrations have systematically increased as a results of fossil fuels use.

If you now apply these distinctions to quantify the probability that a rise in temperature is casual, you pass the test only at the lower level; i.e. there is a probability higher than 1 to 5% that the increase is not due to external causes. But the test depends very much on which question you ask. For instance, instead you might ask the question, "Temperature has been changing very rapidly, and how the normal is that? How normal is the accelerated warming from 1980 to 2000?" The answer is a very simple calculation, and you pass the significance test with flying colors. There is less than 1% chance that the rate of change of temperature may be due just to the normal fluctuations in climate. That is one of the reasons why scientists are so confident that what we are seeing here is not only real but also something that is outside the normal habits of the Earth's climate as we know it at least up to now.

The tests heavily depend on the size of the departure, so if the temperature keeps increasing at this pace, in a relatively short time the tests will be passed to everybody's satisfaction. But even then there will be several anomalous aspects that show up when you look at the climate from different angles - like the question of the speed of warming. In fact, most climate researchers are convinced that there is no other explanation for the observed warming than the effect of carbon dioxide and other gases. Basically we are running out of alternative explanations to attribute the warming trend

---

Bettina Menne
It would be very interesting to run a simulation exercise for Naples or for Palermo using the carbon neutral cities approach. Giving them one euro, for example, for each carbon to saved, to sell to somebody else, or giving them three months of life because of air pollution reduction. Playing it with this terminology might be an interesting exercise.

Claudia Bettiol
How can we increase pragmatism in political choice? Political choice can be positive. As a matter of fact, positive change has started in those countries where politicians have done their job, and we can talk about the positive consequences. For example, in Germany the government

Positive change has started in those countries where politicians have done their job

to, but this does not necessarily mean that the story is over.

Science progresses as a sort of collective temporary skepticism. At a certain point in time an agreement can be reached based on available evidence, but new and improved evidence may change our view. Someone may come up with a very smart and exotic explanation of how the warming trend is possible without referring to carbon dioxide, and in a manner consistent with our observations. In the meanwhile, signs of change abound. The Arctic is warming steadily since the 1950s. There is currently 40% less volume of ice in the Arctic then in the 1950s. It is conceivable that very soon we will have an ice-free Arctic during the Northern Hemisphere summer. This situation would imply changes in the shipping routes between Europe and the Far East. It will open the Arctic to exploitation of mineral and biological resources. This may compound the problem of climate change if we find that there are extensive reserves of fossil fuels in the Arctic Ocean.

The absence of an international agreement concerning exploitation of the Arctic will create a complex geopolitical situation, and there is urgent need of an international treaty to define a shared way of operating, exploiting, and protecting the Arctic. The warming of the Northern polar region is also creating some considerable concern about Greenland. In recent years Greenland glaciers have been melting at an increasing rate. The disappearance of the Arctic sea-ice has little consequences for the global sea level because floating ice does not increase the volume of the oceans when it melts, but the land-locked Greenland glaciers are another story. Greenland is a big reservoir of water and the complete melting of its glaciers could result in up to 7 m of increase of Ward Ocean sea level. Nobody expects Greenland to just disappear so rapidly, but there is a degree of concern because we know very little of the dynamic of

Fig 18.1. Ice Cover Change Between the Period 1979-2000 and 2005

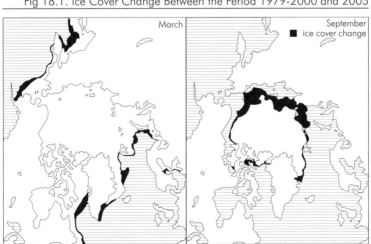

Source: NOAA / National Snow and Ice Data Center (NSIDC)

the ice rivers in Greenland. We know very little of mechanisms that can accelerate or slow down their processes. We do not understand very well the mechanism of accumulation of ice over land. This is something that is now being monitored very closely, and there is a lot of interest.

Once you have accepted the extreme complexity of the climate system, you find yourself in a logical corner. If the system is so complicated how can we understand it? How can we make science with it? This difficulty lies in the sense that science depends heavily on the possibility to perform crucial experiments that enable one to distinguish between different competing theories. Consider, for instance, the famous experiment of Michelson and Morley that confirmed the existence of the electromagnetic ether. The experiment opened the road to spatial relativity; yet the entire experimental apparatus was sitting on a table in a basement!

Now I would very much like to make experiments myself. I would like to close the Strait of Gibraltar, for instance, so that I could check if the Mediterranean Sea is actually drying up at about a meter per year as our estimate indicates. But it is difficult to find the funding and the environmental permission to make such an experiment. So our issue is: how can you make science if you can't perform experiments? How can you have a rational understanding of a realm where no experiments are possible and everything is so complex?

The development of climatology as a quantitative science has been crippled by this problem, such that it basically evolved as a descriptive science in the 19th century and the early part of the 20th century. The situation changed when we achieved the possibility to make experiments. This is possible because we know very well

---

decided to invest 3% of the GDP in photo voltaic energy. Everybody thought they were foolish. Now they are the global leaders in the number of patents, and the number of people working in the field of renewable energy in Germany has exceeded the number of people working in the automobile industry. Toyota is also doing better than GM also thanks to the hybrid engines that they are producing. So the problem now is not what the negative consequences are but what the positive consequences might be. Abu Dhabi and the Arab countries reason that if they sell oil for one hundred dollars, and they sell it for 20 at home they still have 80 dollars available to change their economies. And as the oil-producing countries they are protagonists in this field, it is a question of political choices but also the choices of multinational companies.

So the problem now is not what the negative consequences are but what the positive consequences might be

how the climate system works. We are able to write down the equations that regulate its behavior, but we are unable to solve them, because of their extreme mathematical complexity. The solutions to these equations are possible only through a trade-off. We traded the mathematical complexity of the equations for a large number of elementary operations, using numerical methods and getting approximate solutions. This is what we call a numerical model - a mathematical representation of the functioning of the Earth's climate system.

Present numerical climate models are very advanced. They include not only the physical part of the atmosphere and the ocean, but also the biosphere - the bio-system, the chemistry, the ice - basically we can model everything that is susceptible to mathematical formulation. In the virtual world of the model we can make experiments because we can take out mountains. We can take out the Alps, or not to be stingy, we can take out all mountains and see what the climate of the Earth would be. The models are for us what telescopes are for astronomers, large and sophisticated instruments that allow us to see deeper and deeper into the workings of the climate system. Every new generation of modeling allows us to be more accurate, more detailed, more reliable.

The next-generation model that we are building at the Euro-Mediterranean Center for Climate Change (CMCC) will be able to get climate details that will be useful to the evaluation of climate change impact for a variety of applications that really need localized data. The issue will be, of course, how precise are the simulation data? What are the uncertainties? What level of accuracy can we reach? There is a lot of work to be done, but you need to learn to walk before you start running. The main effect of climate change will be a shifting of precipitation overall, so precipitation may increase

---

The same should have a global outlook; they should realize what opportunities they can grasp in order to change social and economic scenarios. Skipping altogether the debate of climate changes, whether it is real or not.

Luigi Massa

We have built a model of town planning on a specific economic development model and now we have to change this economic model

I believe that political decision makers do not have uncertainties around climate change. The problem is that we have a new chapter of priorities for politicians and economists. We have built a model of town planning on a specific economic development model and now we have to change this economic model. Most people today have not yet understood how to avoid being overwhelmed by consumer goods. We don't yet have this new culture. So today our advertising sells cars - less-polluting cars - but

in some regions and decrease in others. In the Mediterranean, for instance, the winter precipitation is going to decrease quite a bit, indicating a potential serious issue for the next decades in the region that is already under water stress.

So the question is, what can we do? We have discussed mitigation and adaptation as possible options. They are *Wunderwaffen*, and their limitations indicate that we will have to include both in our strategies. The reason we need adaptation is that there is a delay in the increase of carbon dioxide in the climate system and the actual increase in temperature. If by some magic we could limit the accumulation of carbon dioxide to the present level, the earth would still warm up 1.5°C in the next few decades. The warming potential of the carbon dioxide that is already in the atmosphere has not completely expressed itself yet. Adaptation, therefore, is needed, because even in this miraculous scenario, we would still have more global warming. But adaptation also has limitations. There is a limit to the adaptation capacity of cultures and economies and so mitigation strategies that cut down the greenhouse gases emissions are also needed.

There are a lot of international efforts trying to evaluate these strategies and the coming climate changes. For instance, there is a European Community project, CIRCE Climate Change and Impact Research: the Mediterranean Environment, whose goal is to make a regional assessment of climate change and impact in the Mediterranean.[2] It is very interesting since it is one of the first regional assessments of climate change in the international scene. CIRCE launches a new approach for these studies. You usually start with climate drivers. Some scenarios will indicate how much temperature or precipitation is going to change; others will come up with the changes that will occur in agriculture; and the economists will

---

still only cars are being advertised. Alternatives are not being advertised. Public transport systems are not being advertised.

Bettina Menne
When you know what to do, you face the risks. I think the question today lies in making policy. What objectives do we want to achieve? What objectives does my community want to achieve? What objectives do my city want to achieve, or my region, or my country? At the end of the day, we are asking for green house gas reduction. Now.

make the calculation and attribute a monetary figure to the scenario changes.

However, in the middle there is policy. The same set of climate-drivers with a certain policy will give you a certain monetary impact, but if you change the policy then you get another impact, and obviously a different cost. This means that we have to include hypotheses of policy in our assessment of climate change, in our evaluation of climate change impacts, otherwise we won't be able to make a valuation of our options. It is a very challenging problem. And of course this is the reason why things are really difficult.

Drastic reduction in emissions is needed to stabilize the concentrations, far more effective than the 3% reduction that can be achieved under Kyoto protocol. The post-Kyoto climate regime agreement must be effective, as it must be capable of realizing significant emissions reduction. It should be empowering, because it must allow emerging countries to continue their chosen developing paths. It will have to be fair, because it must recognize the historical responsibility of mature economies that have caused the bulk of the problem. It must be shared, because without consensus, both international and domestic, we will never be able to achieve the level of 70 or 80% reductions that are needed.

This is a crucial point. We need a coalition both internationally and domestically, because without a very wide consensus, it will not be possible to implement the policies that are needed. In a 2007 general social survey in the United States, a sample was asked if they were satisfied with their jobs; if they were satisfied with their income; and so on.[3] The results showed that 76% were happy with their income and jobs; 62% expected their position to improve over the next five years. However, only 25% believed that the country

Fig 18.2. The CIRCE Strategy

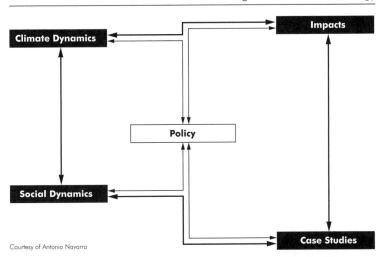

Courtesy of Antonio Navarra

was on the "right track;" 80% thought that the Congress had accomplished nothing; 60% expected that the next generation would be worse off than the current one. This is a very interesting result because it shows that there is a "happiness gap." People are basically satisfied with their lives, but they are not satisfied by what the government is doing to counteract the really large forces that are outside their control: climate change, terrorism, the credit crisis; and they doubt the capacity of the government to handle these problems effectively. The challenge we face is therefore clear: you have to draft policy that is actually tackling these large issues, without touching the personal sphere of satisfaction of the individual. Here is the entire climate change dilemma in a nutshell. This is why it is so controversial to define a policy about climate change, but it is a challenge we have to take if we want to shape a better future for all of us.

1. P. Brohan, J.J. Kennedy, I. Harris, S.F.B. Tett, and P.D. Jones, "Uncertainty Estimates in Regional and Global Observed Temperature Changes: A New Dataset from 1850," *Journal of Geophysical Research* 111, D12106, doi: 10.1029/2005JD006548 (2006).
2. CIRCE, http://www.circeproject.eu
3. D. Brooks, "The Happiness Gap," *New York Times*, Oct. 30, 2007, http://www.nytimes.com/2007/10/30/opinion/30brooks.html (accessed June 2008).

AUTHORS

CINZIA ABBATE

is a founder of the architectural studio AeV in Rome with Carlo Vigevano, and is Clinical Professor and Director of the Roman Studies Program at Rensseleaer Polytechnic Institute. She is the editor of *L'integrazione architettonica del fotovoltaico: esperienze compiute* (The Integration of Photovoltaics in Building: Case Studies), 2002.

LORENZO BELLICINI

is an architect and the Director of the Italian Centre of Economic, Sociological and Market Research for the Construction Industry (CRESME). He is coauthor, with Richard Ingersoll, of *Periferia Italiana* (Italian Periphery), 2001.

HARRIETT BULKELEY

is a Lecturer in Geography at Durham University and Associate Director of the Centre for the Study of Cities and Regions. Her research concerns environmental governance and urban sustainability, with a particular focus on the role of cities in responding to climate change. She is coauthor with Michele Betsill of *Cities and Climate Change: Urban Sustainability and Global Environmental Governance*, 2003.

DAVID BURNEY

is an architect and the Commissioner of Design and Construction for New York City. He was formerly Director of the Design Department at the New York City Housing Authority.

MATTEO CAROLI

is Professor of Economics and Management of International Business at the Università Luiss Guido Carli in Rome, where he directs the PhD Program in Management of Project Financing. His most recent publication is *Un sistema territoriale competitivo* (A Competitive Territorial System), 2005.

ANTONIO CIANCIULLO

is the Environmental Correspondent for the daily newspaper *La Repubblica* in Rome He is the author of *Atti contro Natura* (Acts Against Nature), 1992; *Ecomafia*, 1995; *Far soldi con l'ambiente* (Making Money with the Environment), 1996; *Il grande caldo,* (The Big Heat), 2004; and *Soft Economy*, 2005.

RICHENDA CONNELL

is cofounder and Chief Technology Officer of Acclimatise, a specialist risk management company based in London that assists businesses in adapting to the threats, costs, and opportunities of climate change. She has worked as an environmental consultant to the water and power industries, primarily as an air quality specialist.

LIEVEN DE CAUTER

is Associate Professor in the Department of Architecture at the Katholieke Universiteit Leuven in Belgium. Trained in philosophy, he is an authority on Walter Benjamin, and writes on contemporary art, experience and modernity, and most recently on architecture, the city, and politics. His books include *The Capsular Civilization: On the City in the Age of Fear*, 2004; and he co-edited with Michiel Dehaene, *Heterotopia and the City,* 2008. He was initiator of the Brussels Tribunal on the "Project for the New American Century."

BRUNA DE MARCHI

is an environmental consultant to public institutions, agencies, and companies in Europe, Latin America, and Australia, including the World Health Organization, the Italian Department of Civil Protection, and National Seismic Service. Among her books are *Il rischio ambientale* (Environmental Risk), 2001, coauthored with L. Pellizzoni and D. Ungaro; and *Contribution of Sociology to Disaster Research*, 1987, coauthored with R.R. Dynes and C. Pelanda.

ALESSANDRO LANZA

is the Chief Executive Officer of the ENI Corporate University in Rome. He is the author of *Lo Sviluppo sostenibile* (The Sustainable Environment), 1997; and *Il Cambiamento Climatico* (The Changing Climate), 2000.

BETTINA MENNE

is a medical doctor based in Rome and a specialist in hygiene and public health. In recent years she has coordinated the World Health Organization European Programme on Global Change and Health and several international research and public health projects. She has been a Convening Lead Author of the health chapter and participated in the core writing team of the *Synthesis Report*, IPCC4.

ANTONIO NAVARRA

is a climate scientist with a long international experience in numerical simulation and dynamics. He is the Director of the Euro Mediterranean Center for Climate Change, the Italian organization for climate simulation scenarios.

MATTHEW NISBET

is Assistant Professor in the School of Communication at American University in Washington, DC. His research focuses on the intersection between science, media, and politics. He is the author of numerous research articles on communication, political science, and public policy. He also tracks current events related to strategic communication in his blog "Framing Science" at http://scienceblogs.com/framing-science

JORGE ORTA

is an international artist trained in both fine arts and architecture. He is the founder, with his partner Lucy Orta, of the Studio-Orta in Paris since 1991. He investigates crucial themes of the world today involving community, social linkages, dwelling and habitat, nomadism and mobility, ecology, and sustainable development.

RICHARD PLUNZ

is Professor of Architecture at Columbia University where he is Director of the Urban Design Lab at The Earth Institute and Director of the Urban Design Program at the Graduate School of Architecture, Planning, and Preservation. He is the author of many articles and studies. His books include *A History of Housing in New York City*, 1990; *The Urban Lifeworld: Formation, Perception, Representation*, 2002, coedited with Peter Madsen. A recent publication, coedited with Patricia Culligan, is *Eco-Gowanus: Urban Remediation by Design*, 2007.

CYNTHIA ROSENZWEIG

is a Senior Research Scientist at the NASA Goddard Institute for Space Studies and Columbia Center for Climate System Research. She is a founder of the Urban Climate Change Research Network, and the Chairperson of the Technical Advisory Committee in New York City for the Climate Change Task Force. Her long involvement with climate change research includes the study, *Climate Change and a Global City: The Potential Consequences of Climate Variability and Change — Metro East Coast*, 2001, coedited with William D. Solecki.

MARIANELLA SCLAVI

is Assistant Professor of Urban Ethnography and Cultural Anthropology at the Politecnico di Milano. She is a practitioner of the Consensus Building Approach and of Public Dispute Mediation in Italy. Among her many books are *An Italian Lady Goes to the Bronx*, 2007; *A una spanna da terra* (Six Inches from the Ground) 1989; and *Arte di ascoltare e mondi possibili* (The Art of Listening and Possible Worlds), 2000.

Julie Sze

is Assistant Professor of American Studies and Director of the Environmental Justice Project, John Muir Institute for the Environment at the University of California at Davis. Her research integrates American studies with environmental, urban, and ethnic studies. Her book *Noxious New York: the Racial Politics of Urban Health and Environment*, 2006, analyzes the culture, politics, and history of environmental justice activism in New York City, within the larger context of privatization, deregulation and globalization.

Discussants

Massimo Alesii

is a consultant on community development based in Torino, Italy and former Secretary General of the Fondazione Adriano Olivetti in Rome.

Claudia Bettiol

is Adjunct Professor in the Department of Business Engineering, Università di Roma "Tor Vergata," and a Board Member of the Italian National Agency for New Technologies, Energy and the Environment (ENEA). She has written extensively on environmental technology, and her recent book is *Cuore e ambiente. Passione e razionalità* (Dreams and Environment. Passion and Realities), 2008.

Domenico Cecchini

is an architect and planner and Professor of Urbanism at the School of Engineering of the Università di Roma "La Sapienza." He was former Deputy Mayor for Planning for Rome. He is the editor of numerous articles and books including *Rifare città, studi per ricostruire un quartiere di Roma* (Remaking the City: Studies for Rebuilding a Neighborhood in Rome), 2005. His most recent research focuses on effective sustainability interventions in European urban centers.

Bruna Esposito

is an international artist based in Rome, whose work focuses on the sensorial presence of immaterial, everyday phenomena. Her work has been exhibited at Documenta Kassel, the 48th and 51st Venice Biennale, the Istanbul Biennale, and the Korean Biennale.

Luigi Massa

is former Mayor of Montanaro, Italy. Since 2001 he has served as the City Manager of Napoli, Italy. He is an authority on administrative law and the author of several books including *Sicurezza urbana* (Urban Security), 1999; and *Mine vaganti* (Time Bomb), 2003.

BARTOLOMEO PIETROMARCHI

is an art curator, and former Director of the Art and Society Program, at the Fondazione Adriano Olivetti, Rome.

MARIA PAOLA SUTTO

is a biologist, an environmentalist and journalist based in New York City. She has researched and written extensively on new frontiers of science and on the digital revolution for the Italian media.

GIUSEPPE TRIPALDI

is a biologist and the Director of Ambiente e Territorio (AeT, Agency for the Environment) within the Camera di Commercio Industria Artigianato e Agricoltura di Roma.

BIBLIOGRAPHY

A Green Los Angeles 2006: Recommendations to the City of Los Angeles, a Working Group for a Just and Sustainable Future.

Alonso, J. L., and Mendez, R., 2000: *Innovación, Pequeña Impresa y Desarrollo Local en España*, Civitas, Madrid.

Aalst, Van K. M.,Cannon, M.K., and Burton, T.I., 2007: "Community level adaptation to climate change: The potential role of participatory community risk assessment." *Global Environmental Change*, 18: 165-179.

Beck, U., and Willms, J., 2004: *Conversations with Ulrich Beck*, Polity Press, Cambridge, UK.

Betsill, M., and Bulkeley, H., 2008: "Looking Back and Thinking Ahead: A Decade of Cities and Climate Change Research," *Local Environment: the International Journal of Justice and Sustainability*, 12 (5): 447-456.

Brohan, P., Kennedy, J.J., Harris, I., Tett, S.F.B., and Jones, P.D., 2006: "Uncertainty Estimates in Regional and Global Observed Temperature Changes: A New Dataset from 1850," *Journal Geophysical Research* 111, D12106, doi: 10.1029/2005JD006548

Brooks, D., "The Happiness Gap," *New York Times*, Oct. 30, 2007, http://www.nytimes.com/2007/10/30/opinion/30brooks.html.

Bulkeley, H., and Betsill, M., 2003: *Cities and Climate Change: Urban Sustainability and Global Environmental Governance*, Routledge, London and New York.

Bulkeley, H. and Kern, K., 2006: "Local Government and Climate Change Governance in the UK and Germany," *Urban Studies*, 43 (12): 2237-2259.

Camagni, R., 2002: "On the Concept of Territorial Competitiveness: Sound or Misleading?" *Urban Studies*, 39 (13): 2395-2411.

Camagni, R., and Capello, R., 2005: "ICTs and Territorial Competitiveness in the Era of Internet," *The Annals of Regional Science*, Springer-Verlag.

Campanella, T.J., 2008: *The Concrete Dragon. China's Urban Revolution and What It Means for the World*, Princeton Architectural Press, New York.

Caroli, M., 2005: *Un Sistema Territoriale Competitivo, Capacità Innovativa e Fattori Determinanti Nell'Attrazione Degli Investimenti*, SviluppoLazio, FrancoAngeli, Rome.

Cavill, N., Kahlmeier, S., and Racioppi, F., "Physical Activity and Health in Europe: Evidence for Action," WHO Regional Office for Europe, Copenhagen, 2006, http://www.euro.who.int/document/e89490.pdf.

Center for Science in the Earth System (The Climate Impacts Group), Joint Institute for the Study of Atmosphere and Ocean, University of Washington, and King County, Washington, September 2007: Preparing for Climate Change: A Guidebook for Local, Regional, and State Governments," The Climate Impacts Group, King County, Washington, and ICLEI – Local Government for Sustainability.

Center for Strategic and International Studies, 2007: "The Age of Consequences: The Foreign Policy and National Security Implications of Global Climate Change," Washington, DC., http://www.csis .org/component/option,com_csis_pubs/task,view/id,4154/type,1/ (accessed July 2008).

Chafe, Z., 2007: "Reducing natural disaster risk in cities," *State of the World 2007*, W.W. Norton & Company, New York, London.

Chermayeff, S., 1982: *Design and the Public Good: Selected Writings 1930-1980,* Plunz, R., ed., The MIT Press, Cambridge, MA.

Climate Ark News Archive, June 7, 2007: "Climate change Battle Could Spell New Disasters." http://www.climateark.org/shared/reader /welcome.aspx?linkid=77376&keybold=climate%20change%20disasters (accessed January 2007).

Coafee, J., and Healy, P., 2003: "My Voice, My Place: Tracking Transformations in Urban Governance," *Urban Studies*, 40 (10): 1979-1999.

Cohen, R., and Bell, P., 2007: "National Journal Insiders Poll." *National Journal*, February 3, 2007. http://www.nationaljournal.com/njmagazine/ nj_20070203_2.php.

Colten, C., 2005: *An Unnatural Metropolis: Wresting New Orleans from Nature*, Louisiana State University Press, Baton Rouge.

Colten, C., and De Marchi, B., 2008: "Hurricane Katrina: The Highly Anticipated Surprise," *Città Salute e Sicurezza: Alcuni Riferimenti per il Governo delle aree Urbane*, Menoni, S.,ed., Città Studi, Milano (forthcoming).

Constantin, D., 2006: "Recent Advances in Territorial Competition and Competitiveness Analysis," *Romanian Journal of European Affairs*, 6 (3): October 2006.

Conti, S., and Giaccaria P., 2001: "Economic Development and Local Development," in *Local Development and Competitiveness*, Kluwer Academic Publishers, Netherlands.

Corò, G., 2006: *Crescita, Convergenza e Innovazione: Una Discussione sul Modello di Sviluppo Regionale Nell'UE, Nota di Lavoro*, Dipartimento di Scienze Economiche, www.dse.unive.it/pubblicazioni/ (accessed July 2008).

Cunningham, S., 2002: *The Restoration Economy: The Greatest New Growth Frontier*, Berrett-Koehler Publishers, San Francisco, CA.

Cuzzolaro, M. and Frighe, L., 1991: *Reazioni umane alle catastrofi*, Gangemi Editore.

Daley, H.E., and Farley, J., 2004: *Ecological Economics: Principles and Application*, Island Press, Washington DC.

Davis, M., 2006: *Planet of Slums*, Verso, London.

De Cauter, L., and Dehaene, M., 2007 : "Meditations on Razor Wire : A Plea for Para-Architecture," *Visionary Power: Producing the Contemporary City* (cat.), International Architecture Biennale Rotterdam, NAI Publishers, Rotterdam, Netherlands. 233-247.

De Cauter, L., 2004: *The Capsular Civilization: On the City in the Age of Fear*, NAI Publishers, Rotterdam, Netherlands.

De Cauter, L., and Dehaene, M., 2007: *The Archipelago and the Ubiquitous Periphery: Snapshots of Disaster City*, Air de Paris, Centre Pompidou, Editions du Centre Pompidou, Paris, 144-147.

Dehaene, M., De Cauter, L., eds, 2008: *Heterotopia and the City. Public Space in a Postcivil Society*, Routledge, London and New York.

De Marchi, B., 2007: "Not Just a Matter of Knowledge. The Katrina Debacle," *Environmental Hazards*, 141-149.

Department of City Planning, Population Division, 2006: "New York City Population Projections by Age/Sex and Borough, 2000-2030," New York.

Diamond, J., 2005: *Collapse: How Societies Choose to Fail or Succeed*, Viking, New York

Dominguez, V., 2005: "Seeing and Not Seeing: Complicity in Surprise," http://understandingkatrina.ssrc.org/Dominguez/ (accessed July 2008).

Dowden, M., and Marks, A.C., 2005: "Come Rain or Shine," *Estates Gazette*, July 16.

Dunlap, R. E., 2008: "Climate-Change Views: Republican-Democratic Gaps Expand." Gallup News Service, May 29, 2008. http://www.gallup.com/poll/107569/ClimateChange-Views -RepublicanDemocratic-Gaps-Expand.aspx.

EEA (European Environment Agency) 2001: *Late Lessons from Early Warnings: The Precautionary Principle 1896-2000*, Office for Official Publications of the European Communities, Luxembourg, http://reports.eea.europa.eu/environmental_issue_report_2001_22/en (accessed July 2008).

EU Commission 2007: *Adapting to Climate Change in Europe – Options for EU Action*, Green Paper from the Commission to the Council, the European Parliament, the European Economic and Social Committee and the Committee of the Regions,{SEC(2007) 849} /* COM/2007/0354 final */

European Commission 2007: *Local Energy Action – EU Good Practice*, http://www.managenergy.net/download/local_energy_action_2007.pdf (accessed August 2008).

European Commission 2007: *The Burden of Crime in the EU: A Comparative Analysis of the European Crime and Safety Survey (EU ICS) 2005*, http://www.crimereduction.homeoffice.gov.uk/statistics/statistics060 .htm (accessed July 2008).

Fisher, R., Ury, W., and Patton, B., 1991: *Getting to Yes: Negotiating Agreement Without Giving In*, Penguin Books, New York.

Foroohar, R., 2006: "Unlikely Boomtowns," *Newsweek International*, Jul. 3.

Forum for Urban Design, 2007: "A critical situation: What to Make of Starchitecture and Who to Blame (or Credit)," New York.

Funtowicz, S., and Strand, R., 2007: "Models of Science and Policy," *Biosafety First*, T. Traavik & Lim Li Ching, eds., Tapir Academic Press, Trondheim.

Garderner, B., Martin, R., and Tyler, P., 2003: *Competitiveness, Productivity and Economic Growth across the European Regions*, Cambridge University Press, Cambridge.

Gill, S., Handley, J., Ennos, R., and Pauleit, S., 2007: "Adapting Cities for Climate Change: the Role of the Green Infrastructure," *Built Environment*, 33 (1): 115–133.

Global Green USA, 2007: http://www.globalgreen.org/greenbuilding/ katrina.html (accessed December 2007).

Gornitz, V., Horton, R., Siebert, A., and Rosenzweig, C., 2006: "Vulnerability of New York City to storms and sea level rise." *Geol. Soc. Amer. Abstr. Programs*, 38, no. 7, 335.

Graham, S., and Marvin, S., 2001: *Splintering Urbanism: Networked Infrastructures, Technological Mobilities and the Urban Condition*, Routledge, London.

Greater London Authority, 2007: Action today to protect tomorrow: The Mayor's Climate Change Action Plan, http://www.london.gov.uk/mayor/environment/climate-change/docs/ccap_fullreport.pdf (accessed July 2008).

Gregovius, F., 1968: *Passeggiate per l'Italia*. Avanzini e Torraca, Rome. Haag, A.L., 2007: "Is This What the World's Coming to*?" Nature, Reports: Climate Change.*

Habermas, J., 1987: *The Theory of Communicative Action, Volume Two: The Critique of Functionalist Reason,* Beacon Press, Boston.

Habermas, J., 1990: *Moral Consciousness and Communicative Action,* The MIT Press, Cambridge.

Hammer, S., 2008: "Renewable energy policymaking in New York and London: Lessons for other 'world cities'?" Forthcoming in *Urban Energy Transition: From Fossil Fuels to Renewable Energy*, Elsevier Press, Oxford, UK, (forthcoming).

Harmes – Liedtke, U., 2007: *Benchmarking Territorial Competitiveness*, Working Paper, September 2007, http://www.mesopartner.com/publications/mp-wp9_Benchmarking.pdf (accessed July 2008).

Heal, G., 2000: *Nature and the Marketplace: Capturing the Value of Ecosystem Services*, Island Press, Washington, DC.

Henderson Global Investors 2008: *Managing the Unavoidable: Understanding the Investment Implications of Adapting to Climate Change.* Universities Superannuation Scheme, RAILPEN Investments, Insight Investment, http://www.insightinvestment.com/Documents/responsibility/Reports/Managing_the_Unavoidable_Understanding_the_investment_implications_of_adapting_to_climate_change.pdf (accessed July 2008).

Holling, C.S., 1973: "Resilience and Stability of Ecological Systems," *Annual Review of Ecology and Systematics*, 4: 1-23.

IMF 2007: "Climate Change and the Global Economy," *IMF Survey Magazine*, IMF Research, October 26, 2007.

IMF 2007: "World Economic Outlook. Globalization and Inequality," http://www.imf.org/external/pubs/ft/weo/2007/02/index.htm

Innes, J.E., et al., 2006: *Collaborative Governance in the CALFED Program: Adaptive Policy Making for California Water*, Working Paper # 2006-01, Institute of Urban and Regional Development, University of California, Berkeley, CA, January 2006, http://repositories.cdlib.org/cgi/viewcontent.cgi?article=1044&context=iurd (accessed July 2008).

Innes, J.E., Booher, D.E., 2005: "Reframing Public Participation: Strategies for the 21st Century," *Planning Theory and Practice*, 5 (4): 419-436.

IPCC 2007: *Climate Change 2007: Synthesis Report. Contribution of Working Groups I, II and III to the Fourth Assessment Report of the Intergovernmental Panel on Climate Change*. Core Writing Team, Pachauri, R.K and Reisinger, A., eds., IPCC, Geneva, Switzerland.

IPCC 2007: *Climate Change 2007: The Physical Science Basis. Contribution of Working Group I to the Fourth Assessment Report of the Intergovernmental Panel on Climate Change*, Solomon, S., Qin D., Manning, M., Chen, Z., Marquis, M., Averyt, B.K., Tignor, M., and Miller, L.H. , eds., Cambridge University Press, Cambridge, and New York.

IPCC 2007: *Climate Change 2007: Impacts, Adaptation and Vulnerability. Contribution of Working Group II to the Fourth Assessment Report of the Intergovernmental Panel on Climate Change*, Parry, L.M., Canziani, F.O., Palutikof, P.J., van der Linden, J.P., and Hanson, E.C., eds., Cambridge University Press, Cambridge.

IPCC 2007: *Climate Change 2007: Mitigation. Contribution of Working Group III to the Fourth Assessment Report of the Intergovernmental Panel on Climate Change*, B. Metz, B., O.R. Davidson, R.O., Bosch, R.P., Dave, R., Meyer, A.L., eds., Cambridge University Press, Cambridge and New York.

Isaacs, W., 1999: *Dialogue and the Art of Thinking Together*, Doubleday, New York.

Kennedy, D., 2007: "Breakthroughs of the Year 2007," *Science* 318: 5858, 1833.

Kitson, M., Martin, R., and Tyler, P., 2004: *The Regional Competitiveness Debate*, Competitiveness Report, Cambridge-MIT Institute, Cambridge, UK.

Klein, N., 2007: *The Shock Doctrine: The Rise of Disaster Capitalism*, Metropolitan Books, London, New York.

Klostermann, J.E.M., and Tukker, A., 1998: *Product Innovation and Eco-efficiency*, Kluwer Academic Publishers, Norwell, MA.

Kolbert, E., 2006: *Field Notes from a Catastrophe: Man, Nature, and Climate Change*, Bloomsbury USA, New York.

Kraft, J.C., et al.: "Paleographic Reconstruction of Coastal Aegean Archeology Sites," *Science*, 195, 1977.

Kolk, A., 2000: *Economics of Environmental Management*, Pearson Education Limited, Harlow.

Kopk, J.T.M., and de Coninck, C.H., 2007: "Widening the scope of policies to address climate change: Directions for mainstreaming," *Environmental Science & Policy*: 587-599.

Krugman, P. R., 1996: "Making Sense of the Competitiveness Debate," *Oxford Review of Economic Policy*, vol. 12, n. 3.

Krugman, P., 1994: *Il Mito del Miracolo Asiatico*, Internazionale Mondatori, Milan.

Krugman, P., 1997: *Pop Internationalism*, The MIT Press, Cambridge, MA. Linden, E., 2006: "Cloudy with a Chance of Chaos," *Fortune*, Feb. 27: 135-145.

London Leading to a Green London: Green homes concierge service, http://www.greenhomesconcierge.co.uk/ (accessed July 2008).

Lowndes, V., 2001: "Rescuing Aunt Sally: Taking Institutional Theory Seriously in Urban Politics," *Urban Studies*, 38 (11): 1953-1971.

Lyman, P., and Varian, H.R., 2003: *How Much Information 2003?* University of California, Berkeley, CA, http://www2.sims.berkeley.edu/research/ projects/how-much-info-2003/.

Martin, R., 2005: *Thinking about Regional Competitiveness: Critical Issues,* Cambridge-MIT Institute, Cambridge, UK.

Mayor's Office of the City of New York 2007: PlaNYC 2030, New York, New York. http://www.nyc.gov/html/planyc2030/html/home/ home.shtml, http://www.nyc.gov/html/planyc2030/downloads/pdf/ progress_2008_climate_change.pdf (accessed July 2008).

Mayor's Office of the City of New York 2008: PlaNYC Progress Report, New York.

McFadden, L., Nicholls, R., and Penning-Rowsell, E., 2007: *Managing Coastal Vulnerability,* Elsevier, Oxford and Amsterdam.

McKibben, B., 1989: *The End of Nature*, Random House, New York.

Meadows, D.H., et al., 1972: The *Limits to Growth: A Report for the Club of Rom's Project on the Predicament of Mankind*, Universe Books, New York, http://www.clubofrome.org/docs/limits.rtf (accessed January 2008).

Menne, B., and Ebi, K.L., 2006: *Climate Change and Adaptation Strategies for Human Health*, Steinkopff Verlag, Darmstadt.

Mileti, D., 1999: *Disasters by Design: A Reassessment of Natural Hazards in the United States,* Joseph Henry Press, Washington, DC.

Mishan, E.J., 1967: *The Cost of Economic Growth*, Praeger Publisher, New York.

Monstadt, J., 2007: "Urban Governance and the Transition of Energy Systems: Institutional Change and Shifting Energy and Climate Policies in Berlin," *International Journal of Urban and Regional Research*, 31 (2): 326-

343.

New York City Independent Budget Office, 2005: Understanding New York's Budget: A Guide, http://www.ibo.nyc.ny.us/ (accessed July 2008).

New York State Office of the State Comptroller, 2007: New York's Public Authorities, http://www.osc.state.ny.us/pubauth/index.htm (accessed July 2008).

Newell, P., 2005: "Race, Class and the Global Politics of Environmental Inequality," *Global Environmental Politics* 5: 70-94.

Nisbet, M., and Myers, T., 2007: *The Polls—Trends*: "Twenty Years of Public Opinion about Global Warming," *Public Opinion Quarterly*, 71 (3): Fall, 444-470.

Nordhaus, T., and Shellenberger, M., 2007: *Break Through: From the Death of Environmentalism to the Politics of Possibility*, Houghton Mifflin, Boston, MA.

Nordhaus, T., and Shellenberger, M., 2007: "Second life: A manifesto for a new environmentalism." *The New Republic*, September 24, 31-33

Norwegian Institute for Urban and Regional Research (NIBR), http://www.nibr.no/content/view/full/66 (accessed August 2008).

OECD 2000: *The Contribution of Environmental Management Systems to the Establishment of Territorial Development Policies*, Discussion Paper, Territorial Development Service, OECD, Geneva.

Offutt, A. J., 1973: "Meanwhile, We Eliminate," in Elwood, R., ed., *Future City*, Trident Press, New York.

Ourousoff, N., 2008: "City on the Gulf: Koolhaas Lays Out a Grand Urban Experiment in Dubai," *The New York Times Architectural Review*, March 3, 2008. http://www.nytimes.com/2008/03/03/arts/design/03kool.html

Pan-European Program, "Transport, Health and Environment Pan-European Programme," United Nations Economic Commission for Europe/WHO Regional Office for Europe, Geneva/Copenhagen, 2008, http://www.thepep.org.

Panitch, L., and Leys, C., 2006: *Coming to Terms with Nature: Socialist Register 2007*, The Merlin Press, London.

Perrow, C., 1999: *Normal Accidents: Living with High Risk Technologies*, Princeton University Press, Princeton, New Jersey.

Pew Research Center for the People and the Press, 2008: "A Deeper Partisan Divide on Global Warming," 2008, http://peoplepress.org/report/417/a-deeper-partisan-divide-over

-global-warming.

Pielke, A.R. Jr., and Sarewitz, D., 2005: "Bringing Society Back into the Climate Debate," *Population and Environment*, 26 (3): 255-268.

Pielke A.R. Jr., et al., 2007: "Lifting the taboo on adaptation." *Nature*, 445: 597-598.

Piven, F.F., 2004: *The War at Home: The Domestic Costs of Bush's Militarism*, The New Press, New York and London.

Porter, M.E., and Van Der Linde, C., 1995: "Green and Competitive: Ending the Stalemate," *Harvard Business Review*, Sept.-Oct.

Portes, A., 1998: "Social Capital: Its Origins and Application in Contemporary Strategy," in *Annual Review of Sociology*, 24.
Pulido, L., and Peña, D.G., 1998: "Environmentalism and Positionality: The Early Pesticide Campaign of the United Farm Workers' Organizing Committee, 1965-1971," *Race, Gender & Class* 6: 33-50.

Putt del Pino, S., Levinson, R., Larsen, J.: *Hot Climate, Cool Commerce: A Service Sector Guide to Greenhouse Gas Management*, World Research Institute, Washington, DC.

Racioppi, F., et al., 2008: "Preventing road traffic injury: a public health perspective for Europe," WHO Regional Office for Europe, Copenhagen, http://www.euro.who.int/document/E82659.pdf.

Ranghieri, F., Sinha, R., Kessler, E., 2008: *Climate Resilient Cities: A Primer on Reducing Vulnerabilities to Climate Change Impacts and Strenghtnening Disaster Risk Management East Asian Cities,* The World Bank, Washington, DC.

Rizzi, P., 2007: *Sviluppo Locale e Capitale Sociale. Il Caso delle Regioni Italiane,* Working Paper.

Rosenzweig, C., Solecki, W., eds., 2001: *Climate Change and a Global City: The Potential Consequences of Climate Variability and Change — Metro East Coast.* Report for the U.S. Global Change Research Program, National Assessment of the Potential Consequences of Climate Variability and Change for the United States. Columbia Earth Institute, Columbia University, New York.

Rosenzweig, C., Gaffin, S., and L. Parshall, eds., 2006: Green Roofs in the New York Metropolitan Region: Research Report. Columbia University Center for Climate Systems Research and NASA Goddard Institute for Space Studies.

Rullani, E., 1985: "Territorio e Informazione: i sistemi locali come forma di organizzazione della complessità," *Economia e Politica Industriale*, 45.

Rullani, E., 1997: "Economia globale e diversità," *Economia e Politica Industriale*, 94.

Sassen, S., 2007: *The Endless City*, Phaidon Press Inc., New York.

Schienstock, G., Kautonen, M., and Roponen, P., 2003: *Cooperation and Innovation as Factors of Regional Competitiveness: A Comparative Study of Eight European Region*, http://www.geo.ut.ee/nbc/paper/schienstock _kautonen.htm_(accessed July 2008).

Schlosberg, D., 2004: "Reconceiving Environmental Justice: Global Movements and Political Theories," *Environmental Politics,* 13: 517-540.

Schmidheiny, S., ed., 1992: *Cambiare Rotta – Una Prospettiva Globale del Mondo Economico Industriale sullo Sviluppo e l'Ambiente,* Il Mulino, Bologna.

Segre, A., and Dansero, E., 1996: *Politiche per l'Ambiente – Dalla Natura al Territorio,* UTET, Turin.

Sepic, D., 2005: *The Regional Competitiveness: Some Notions,* Russian European center for Economic Policy (RECEP), Moscow. http://

www.recep.ru/files/documents/regional_competitiveness_en.pdf (accessed July 2008).

Seri, P., 2001: "Losing Areas and Shared Mental Models: Towards a Definition of the Cognitive Obstacles to Local Development*,"* Max Planck Institute for Economic Systems, Gennaio, Germany.

Shaw, R., Colley, M., and Connell, R., 2007: *Climate Change Adaptation by Design: A Guide for Sustainable Communities,* TCPA, London, http:// www.tcpa.org.uk/downloads/20070523_CCA_lowres.pdf (accessed July 2008).

Sinding, K., 2000: "Environmental Management Beyond the Boundaries of the Firm: Definitions and Costraints," *Business Strategy and the Environment,* 9 (2).

Singapore Environment Institute 2005: The Worldwide Sustainability Timeline, National Environment Institute, Singapore, http://www.nea.gov.sg/cms/sei/SEIsustainabilitytimeline.pdf (accessed June 2008).

Sloterdijk, P., 1989: *Eurotaoismus: Zur Kritik der Politischen Kinetik,* Suhrkamp, Frankfurt.

Smith, K.R., Mehta, S., and Maeusezahl-Fuez, M., 2004: "Indoor Air Pollution from Household Use of Solid Fuels," M. Ezzati et al., eds., *Comparative Quantification of Health Risks: Global and Regional Burden of Disease Attributable to Selected Major Risk Factors,* World Health Organization, Geneva, 1436–1493.

Smith, N., 2006: *Coming to Terms with Nature: Socialist Register 2007,* The Merlin Press, London.

Soja, E., and Kanai, M., 2007: "The urbanization of the world," *The Endless City,* Phaidon Press Inc., New York.

Solecki, W.D., Rosenzweig, C., 2004: "Biodiversity, biosphere reserves, and the Big Apple: A study of the New York Metropolitan Region," *Ann. New York Acad. Sci.,* 1023, 105-124, doi:10.1196/annals.1319.004.

State of the News Media, 2008: http://www.stateofthenewsmedia .org/2008/narrative_overview_contentanalysis.php?cat=2&media=1.

Stern, N., 2006: *The Economics of Climate Change: The Stern Review,* HM Treasury and the Cabinet Office, London, http://www.hm-treasury.gov .uk/independent_reviews/stern_review_economics_climate_change/ sternreview_index.cfm (accessed February 2008).

Susskind, L., and Cruikshank, J., 1987: *Breaking the Impasse: Consensual Approach to Resolving Public Disputes,* Basic Books, New York.

Susskind, L., and Weinstein, A., 1980: "Toward a Theory of Environmental Dispute Resolution," *Environmental Affairs Law Review,* Boston College, 9.2 (1980):143-196.

Sze, J., 2005: "Katrina: Perspectives from the Social Sciences," Social Sciences Research Council, http://understandingkatrina.ssrc.org/Sze/ (accessed July 2008).

Sze, J., 2007: *Noxious New York: The Racial Politics of Urban Health and Environmental Justice,* The MIT Press, Cambridge, MA.

Tadini, M., 2006: "Dotazioni e Performance dei Sistemi Territoriali: L'analisi delle Province del Nord-Ovest," Dipartimento di Studi per L'Impresa e il Territorio, Working Paper n.12, March 2006.

The Clean Air Partnership 2007: Cities Preparing for Climate Change: A Study of Six Urban Regions, May 2007.

Travaglio, M., 2006: *The Disappearance of Facts.* Il Saggiatore, Milan.

U.S. Congress 2006: A Failure of Initiative - Report by the Select Bipartisan Committee to Investigate the Preparation for and Response to Hurricane Katrina, U.S. Government Printing Office, Washington, http://www.gpoaccess.gov/katrinareport/mainreport.pdf (accessed July 2008).

UK Climate Impacts Programme 2008: http://www.ukcip.org.uk/ (accessed July 2008).

UNEP (United Nations Environment Programme) 1992: *Rio Declaration on Environment and Development,* http://www.unep.org/Documents .Multilingual/Default.asp?DocumentID=78&ArticleID=1163 (accessed July 2008).

UNFCCC 2005: *Vulnerability and Adaptation to Climate Change in Europe*, EEA, Technical Report No 7/2005, ISSN 1725-2237, http://reports.eea .europa.eu/technical_report_2005_1207_144937/en/EEA_Technical_ report_7_2005.pdf (accessed July 2008).
UNFCCC 2008: *Nairobi Work Programme on Impacts, Vulnerability and Adaptation to Climate Change*, Nairobi Work Programme, http://unfccc .int/adaptation/sbsta_agenda_item_adaptation/items/3633.php (accessed July 2008).

UNFCCC 2008: *Decision-/CP.13, Bali Action Plan*, Advanced Unedited Version, http://unfccc.int/files/meetings/cop_13/application/pdf/ cp_bali_action.pdf (accessed July 2008).

United Nations 2007/2008: *Fighting Climate Change: Human Solidarity in a Divided World*, Human Development Report, http://hdr.undp.org/en/ media/hdr_20072008_en_complete.pdf (accessed July 2008).

Urban Climate Change Research Network 2008: http://www.uccrn.org/ Site/Home.html (accessed July 2008).

Urban Design Lab 2008: http://www.urbandesignlab.columbia.edu (accessed August 2008).

Vaccà, S., 1983: "L'ambiente come Forza Produttiva," *Politica ed Economia*, 10.

Varaldo R., and Caroli, M.G., 1999: "Il Marketing del Territorio: Ipotesi di un Percorso di Ricerca," *Sinergie*, 49, maggio-agosto.

Varaldo, R. 1995: "Dall'impresa Localizzata All'impresa Radicata," *Sinergie*, 36/37: 27-44.

Vogel, C., Moser, C.S., Kasperson, E.R., and Dabelko, D. G., 2007: "Linking Vulnerability, Adaptation, and Resilience Science to Practice: Pathways, Players, and Partnerships," *Global Environmental Change* 17: 349- 364.

Wackernagel, M., and Rees, W., 1994: *Our Ecological Footprint: Reducing Human Impact on the Earth*, New Society Publishers, Gabriola island, BC, Canada.

WALCYING: *How to Enhance WALking and CYcliNG Instead of Shorter Car Trips and to Make These Modes Safer*, Final Project Report, Office for Official Publications of the European Communities, Luxembourg, 1997. http://cordis.europa.eu/transport/src/walcyngrep.htm

WBCSD (World Commission on Environment and Development) 1993: *Eco-efficiency*, WBCSD Report, Geneva.

WCED (World Commission on Environment and Development) 1987: *Our Common Future*, The Brundtland Report, Oxford University Press,

Oxford.

Welford, R., and Gouldson, A., 1993: *Environmental Management and Business Strategy*, Pitman Pub., London.

White House 2003: "President Bush Meets with Prime Minister Blair: Remarks by the President and British Prime Minister Tony Blair," *The Cross Hall*, Nov. 20, http://www.whitehouse.gov/news/releases/2003/01/20030131-23.html (accessed June 2005).

World Economic Forum 2007: The Global Competitiveness Report, 2006-2007, www.weforum.org (accessed July 2008).

World Energy Outlook 2008: The International Energy Agency.

World Health Organization 2006: "Health Impact of PM_{10} and Ozone in 13 Italian Cities," World Health Organization, Copenhagen, Denmark. http://www.euro.who.int/document/e88700.pdf (accessed July 2008).

World Health Organization 2008: Heat Health Action Plans, WHO Regional Office for Europe, Copenhagen, Denmark.

World Health Organization 2008: Protecting Health in Europe from Climate Change, WHO Regional Office for Europe, Copenhagen, Denmark.

The United Nations Population Division: *World Urbanization Prospects:The 2007 Revision Population database*, New York, 2007.

Worldwatch Institute 2007: Report on Progress toward a Sustainable Society, State of the World, Our Urban Future.

Worster, D., 1993: *The Wealth of Nature: Environmental History and the Ecological Imagination*, Oxford University Press.

Wurman, R.S., 1989: *Information Anxiety,* Doubleday, New York.

Yankelovich, D., 1991: *Coming to Judgement: Making Democracy Work in a Complex World*, Syracuse University Press, Syracuse, New York.

Yankelovich, D., 2001: *The Magic of Dialogue*, Touchstone Books, New York, NY.

Young, C.W., and Rikhardsson, P.M., 1996: Environmental Performance Indicators for Business, *Eco-management and Auditing*, 3 (3): November 1996.

Zadek, S., 1999: "Stalking Sustainability," *Greener Management International*, 26.

Zangger, E., 2001: *The Future of the Past: Archaeology in the 21st Century*, Weidenfeld & Nicolson Limited, London.